MW01515562

A Revolution In Understanding

Discovering your natural intelligence

Book II of Decoding Potential

Robert J. Flower, Ph.D.

A Revolution
In
Understanding

Discovering your natural intelligence

Book II of Decoding Potential

Robert J. Flower, Ph.D.

Patent pending on numerous segments
Printer: Central Plains Book Manufacturing
Publisher: Gilchrist Institute
Book design: Patricia Rasch.

ISBN: 0-9759501-1-8

Printed in the United States of America

10 9 8 7 6 5 4 3 2 1

Table of Contents

Introduction

Book I of *Decoding Potential* started with the axiom "When we die, God will ask us just one question: 'What did you do with what I gave you?'" *A Revolution in Understanding: Discovering Your Natural Intelligence* is better described by the famous German poet Johann Wolfgang Von Goethe, "To be pleased with one's limit is a wretched state".

This sequel to Book I of *Decoding Potential: Pathways to Understanding,* is a fascinating, groundbreaking book that is the result, not only of a great deal of study, but also of an intense passion for mind sciences. The systems and science of Natural Intelligence & Thinking were born in 1980 when I had an extended epiphany about the nature of the universe, and hence, an understanding of its dynamics. From that point on, I dedicated myself to a worldwide search for answers, a journey that would have made Indiana Jones jealous.

My studies included sacred sites, the holy scriptures of many religions, physics, biology, philosophy, and theology, all of which describe to greater or lesser degrees the mysterious components of nature. I began to ask myself many of the questions we all ponder from time to time: "What and who are we? Is there a plan to life and the universe? If so, what is it?" Then, one fateful day in 1984, I met geneticist Dr. Derald Langham, who shared some of his own answers to these timeless questions. Dr. Langham had contacted me after reading a financial magazine featuring an article I had written, entitled "The Master Pattern: The Concept of Living Geometry and the Stock Market." Dr. Langham told me that he had identified the geometric structure of the development of a plant cell. We found common ground for what appeared, at first, to be two vastly different subjects. Dr. Langham shared with me the fact that he had also experienced an epiphany at the age of five, an epiphany wherein he had a vision of this geometric form. (An epiphany is defined as a moment of sudden intuitive understanding or a flash of insight. In rare instances, an epiphany can last for months.) Dr. Langham's epiphany was an actual vision of the geometric, mathematical model of the basic cell of a plant. This was later to be described more accurately as a "living geometry." This phrase seemed apt since it was later discovered that this geometry was the basic structure of all creation! His discovery embraced not only the ultimate shape of the plant cell, but also all of its various stages of development. According to Dr. Langham, from the moment of his epiphany, even though he had been only five years old, he knew he was going to become a biologist, or more specifically, a geneticist. He was later to be decorated by Venezuela for his work during World War II for assisting its government in fruit and plant development. It was not long before this brilliant man became an expert in plant genetics.

These shapes and stages of geometry, I realized, were mirrored in all of nature—in living organisms, the dynamics of humankind, the design and creation of the great pyramids, Stonehenge, and, yes, even the stock market. In addition, this living geometry was found in the scriptures, mathematics, and ancient calendars.

From my meeting with Derald came a new passion: to prove that our mutual findings were the Holy Grail, so to speak. We both became completely focused on a method to illustrate how this geometry worked and how it could be used in our lives. We spent a great deal of time together over the next several years until Dr. Langham's death in 1994, discussing his theory and why it was not more widely studied or accepted.

At the time, I believed that, although the theory of cellular geometry did not appear to have any immediate applications, it did, in fact, have immense applications in the study of human intelligence. I became convinced that Dr. Langham's problem in communicating this revelation to his colleagues was that the initial language he used to describe his model was too technical. The geometry needed to be explained in a simpler language, one that would be understood outside the scientific world of genetics.

Combining my own work with Dr. Langham's initial understanding and studies, I set out to find and develop this language. My travels and subsequent investigations eventually led me to the language of ancient philosophies, sciences and religions. Consequently, I came to a profound understanding of those teachings and concepts in a way that has rarely been understood. I saw in them a powerful abstract symbolism, a universal language. This symbolism and its attendant language enabled one to get to the heart of things, as well as to discover hidden patterns and information. This eventually led to the formation of the NATI System (Natural Thinking & Intelligence).

For over twenty-five years, I have studied several thousand cases testing the viability of this system, and in virtually every case, the all inclusive character of NATI provided accurate, clear insight about perceptions, awareness, learning, and interpersonal relationships. I found that whenever I lectured on the NATI system, people of all nationalities, religions, and cultures were fascinated with the concept. Now my research is more than a concept; it is fact. It has been proven to work beyond any reasonable doubt. Thus, NATI is not just about intelligence; it is a way to understand the meaning of life through systems and their dynamics. Everything in the universe, including the human species, shares this same living geometry at its core, from a single plant cell to a human being—even to the design of the universe itself. We now know who, what, and why we are here. This book presents these answers.

The meaning of life and the true essence of things are described within certain scientifically defined principles of nature that are contained within the NATI doctrine. Understanding these principles fosters greater clarity of thinking, as well as a "knowing" and acceptance of both the positive and negative occurrences within our world. Learning and understanding these principles enable us to see the entirety and the profoundness of any issue through a simplified process. Using these principles will allow one to generate models for his or her own direction as he or she sees fit. They present a practical framework for categorizing any type of data and, in fact, have been called by users a "perfect truth vehicle," one that forces us to be honest with ourselves, others, and our entire environment.

While Book I dealt with the scientific validity behind the process of decoding potential, Book II is a pragmatic approach to the subject. In this sequel, the reader will find real-life applications for utilizing the principles of nature and the Doctrine of Potential. In *A Revolution in Understanding*, the theme of nature as a life model blooms, and that is where we shall begin in chapter one.

Throughout the book, the reader should be aware of several important basics. Enhanced understanding and greater intelligence are achieved when we deal with facts and absolutes. In other words, when we decode our potential, the objective of achievement is best realized when we base ourselves in undeniable principles! Further, these undeniable principles are best found in nature!

Throughout the book, the reader will find fascinating applications of its contents. One of these, for example, is the experience of looking at the world through the lens of the Natural Thinking and Intelligence Matrix. A simple example is how we look at things through the principle of focus. This will show us the underlying focus of whatever is being presented to us. Once this occurs, we can then utilize the other principles to better understand what is going on and how to address various issues.

Moreover, people have asked me why we talk of *Decoding Potential*? Why not *Decoding Your Potential* or *Achieving Your Potential*? The reason is this: We, and all of reality, are products of a force we identify as potential. It is not only something within us; it represents our very origins. *Decoding Potential* books are about discovering that force and our origins!

Lastly, in Systems Thinking there is an aspect known as emergent properties. These occur when all of the parts are in place. At that point results begin to emerge. This is how the book will develop. Once we have everything in place and defined, the end result will emerge.

Chapter One

Nature: A Life Model

"Say if thou knowest science.
Wherefrom cometh wisdom, and where is understanding located?
Mortals know not its arrangement."

—*Book of Job*

Let's cut right to the chase. Humankind has unsuccessfully searched for meaning, understanding, and direction since the beginning of time. Governments, religions, societies, and cultures have repeatedly failed us time after time. As the quote from the Book of Job clearly states, we just don't know where wisdom originates. Yes, we have our experts, leaders, diviners, and intelligentsia, and yet we still don't know how to find wisdom! There are many reasons for this, such as blocking ourselves from the truth because of fear, ego, self-deception, and ignorance. We fear truth! We put up ego walls to protect our fragile self-images. We deceive ourselves concerning the obvious—our weaknesses and our frailties. We go to our houses of worship and say "Peace and love" and then go to work and social functions and cut each other up. We are subjective, close-minded, and self-centered—and that's on a good day! If you think any of the above are generalizations that don't relate to you, then you may well be on your way to implementing an escape route from the truth.

Background

Reacting to the Industrial Revolution, the poet William Wordsworth wrote the following lines:

> The world is too much with us; late and soon,
> Getting and spending, we lay waste our powers:
> Little we see in Nature that is ours:

For the famous British poet, the key to inner harmony was to see past the thousands of distractions encountered in everyday living and recognize man's inherent connection to nature. Not much has changed since 1806, when Wordsworth penned the above lines. Indeed, it could be argued that the Industrial Revolution and all subsequent technology have served to distance man from the cosmos that gave him birth. In turn, this has profoundly influenced the way we view the world—our paradigm, if you will—and how we regard our roles within its various systems, be they social, economic, educational, political, or many others. In short, given an explosion in information and technology, especially in the present computer age, people feel disconnected, alienated from everything around them. Ironically, we ourselves are the machine makers, and the resulting feelings of alienation we experience are illusions that we allow our conscious minds to entertain.

It is important to note that the underpinnings of the principles of nature as defined by NATI are not new. From antiquity to the present age, philosophers, scientists, and theologians have asserted that God and nature are one and the same. The logical conclusion is that following nature is the surest path to meaning, creativity, and happiness. As best as can be determined, this concept of naturalness and the importance of adhering to natural laws originated in the second and third centuries B.C. with the Stoics Zeno of Citium and Chrysippius, who believed long before the birth of physics or the scientific method that the universe was governed by laws that exhibited strict rationality. Indeed, following natural laws, even those resulting from simple logic or observation, was the basis of principles that governed both human and non-human behavior, as well as human morality. We also see the same concepts mirrored in Plato's World Soul and Aristotle's Immovable Mover, concepts and forces that show nature reduced to an absolute elegance and simplicity and therefore serving as a fit model for human action. One can only marvel at the precision and accuracy of thought exhibited by these early philosophers.

More recently, and refined to an exacting degree, we see the same notion that God and nature are indivisible in theologian Thomas Aquinas' *Summa Theologica*. The same essential belief is echoed in St Augustine's *City of God*. This principle of divine intelligence permeating the natural order of creation is also quite evident in Russell's Global Brain and Teilhard de Chardin's Noosphere. The paradigm shifts evident in all of these philosophies have eventually led to the views of modern day systems thinkers and psychologists such as Heinz Werner, Jean Piaget, Bernard Kaplan, and Seymour Wapner.

As can be clearly seen, there are indeed alternative ways of looking at the world—new paradigms—that do not require us to regard man as alienated from the cosmos. The cosmos *is* man, and man *is* the cosmos.

By the end of this book, we will have come full circle back to this very point, namely, that just as man is in the cosmos, the cosmos is actually—and incredibly—inside the man. This point is related to the fundaments and structure of nature. These fundamentals of nature I have identified as the thirteen intelligences of NATI, plus Polarity, which we shall examine shortly.

The Essence of Nature Is Its Qualities.

I have come to believe that nature's qualities are the soul and spirit of life. I would define soul and spirit as what we are, what we do, and how we do it. What follows is a list of some of the qualities of nature. Keep in mind as you go through this book that NATI is a scientific method that utilizes specific principles of nature to define our soul and spirit, our inner selves, and our essence that emanates from the processes of integration and development. Therefore the pursuit of a life purpose of potential development naturally generates the following natural qualities:

Instinctive:	Inherent process; innate order
Economical:	No wasted time; energy or resources
All-embracing:	Encompasses all possibilities; applicable to anything
Adoptive:	Flexible
Expansive:	The ability to grow; freedom to develop
Balance:	Possesses equal amounts of positive and negative
Engaging:	Interaction breeds results
Harmonious:	Blends and interacts well; effortless
Excellence:	Highest quality

Effective:	It works; is productive
Efficient:	Perfect order; non-resistive
Disciplined:	Restricts over-expansion; structured
Complimentary:	Integrates competing parts, which develops potential further
Communicative:	It relates smoothly; connects
Nurturing:	Supportive

Essentially then, nature is everything; it is all-encompassing. Its qualities are virtuous even though they are influenced by opposition. This seeming contradiction is discussed at length in this book. These qualities, or virtues, of nature are best evidenced when the *development of potential becomes our life purpose!*

The Decline of Procrastination

Despite revolutions in thought and scientific advances, a static paradigm resulted from Isaac Newton's discovery of the laws of motion and gravity.[1] From Newton's work emerged the philosophy of determinism. According to determinism, the universe may be viewed as a big clock set in motion by a divine hand at the beginning of time and then left undisturbed (a powerful symbol of a very static system). From its largest to its smallest motions, material creation could be determined with absolute accuracy by Newton's laws. All of humanity's tragedies and joys were predetermined in this mechanistic scheme of the universe.

Fast forward to the twentieth century, and we encounter theoretical physics and quantum mechanics, both of which have resulted in a profound paradigm shift that extends to mind sciences and the very nature of reality (metaphysics). Indeed, there was a new importance and immediacy for interpreting things since the emerging experimental physics of the twentieth century stated that elementary particles behaved in random, uncontrollable ways that Newtonian physics could not explain. As Lewis Carroll might have said, things were growing "curiouser and curiouser" in a field of endeavor—science—that was supposed to explain nature with Newton's clockwork precision.

Not discounting the work of James Maxwell Clark, Edward Rienmann, Niels Bohr, and others, Albert Einstein was perhaps the most influential scientist in bringing about this new vision of the universe. For Einstein, the emerging paradigm was nothing less than a conversion from personal religion to the cosmic religion of science. Like Spinoza, Einstein observed the universe as part of an impersonal energy, governed by laws that can be known by us but are independent of our thoughts and feelings.[2] While this may seem as if dear old Albert was once again imposing a Wordsworthian division between humans and nature, the exact opposite is true. The existence of a cosmic code—the laws of material reality that are confirmed by experience—is the very bedrock of faith that influences the work of the natural scientist. Put more succinctly, this kind of modern scientist sees in the code of nature the eternal structure of reality, not as imposed by man, *but as written into the very substance of the universe.* The innateness of the cosmic code can otherwise be described as the inherent components of nature that are so well defined by the thirteen intelligences of Natural Thinking & Intelligence.

Modern quantum mechanics therefore offers us a chance to see our connection to the universe in a unique way. If this is true, our relationship with all matter that surrounds us must be reinterpreted.[3] We *are* directly connected with nature, but notice the importance of a point of reference in establishing this connection. We cannot reach the absolute reality of nature (Plato's World

Soul) without changing our point of reference to help us move from the concrete world around us to the abstract notions that are the outgrowths of twenty-first century science. While many of these focuses are extremely complicated, almost everyone will recognize $E=MC^2$, the famous equation that states energy and matter are two manifestations of the same thing.

Energy Is Key.

We have now arrived at the very heart of the nature of reality and one of the most important principles of NATI: *all reality manifests from the static energy of potential, which is the underlying force of the cosmos.* All of creation comes from a realm of infinite potential—a pregnant void, as Taoists proclaim.

As mysterious as quantum mechanics might seem, it elegantly describes the nature of reality . . . and the reality of nature. All mass in us and around us is simply a form of bound energy made possible by (and subject to) conscious observation.[4] That conscious observation is identified as one's "point of reference." *This point is actually the starting place for the development of potential! It is here where our paradigms, our paths, begin!*

The Universality of Humankind

We have had quite a bit to say about nature and reality thus far. It is heady stuff, to be sure. Perhaps pursuing natural laws is somewhat of a creative game that both scientists and philosophers play with nature. The obstacles in such a venture are the limitations of experimental techniques, not to mention the ignorance and fear that always accompany concepts that help formulate new paradigms. (We shall have more to say about these limitations in a moment.) The goals of such pursuits, however, are unlocking physical laws and, even more importantly, *the internal logic* that governs the entire universe.[5] Let us not forget that the internal logic behind the universe *is nature itself* and that our goal in NATI is to model ourselves after nature.

In our pursuits, however, we cannot get lost in misconceptions about these laws. The difference between physical law and social law, for example, is the difference between "thou shalt not" and "thou can not." No one will ever go to jail for violating the law of thermodynamics.[6] However, the idea of a physical law in an absolute world beyond change is worth the above pursuit. To demonstrate the invariance of physical law, one might consider the necessity of the Law of Magnetism and the Law of Gravity. It is in relation to this last factor that symmetry is implied because it doesn't change—it's invariant. That, essentially, is unchangeable, internal logic. This is precisely why physicists always search for symmetry; for they realize that any new invariance implies the existence of an absolute, and to a scientist, the universality of a law is by far its greatest defining characteristic.

So here's our starting point—the invariance factors evident in nature and humanity. By analogy, finding an invariant human factor that permeates all of humankind, even in varying degrees, would correspond with an absolute in the science of mind mechanics. Is there such a factor? Of course, and that factor is *problems*! It is beyond comprehension that any mind on the planet does not encounter problems on a regular basis. While problems may be an issue of relativity (what is a problem to some is inconsequential to others), the category of problems is a universal, absolute, and quite inescapable factor. But here's the catch: only when we accept the absoluteness of problems can we be in a valid position to achieve our potential. Since we are part of nature, nature is, by necessity, our model for converting potential energy into the goals we have in the here and now. We can formulate a metaphysical corollary to

Heisenberg's Uncertainty Principle by saying that we are mired in uncertainty and problems. How we choose to view those problems, as well as possible solutions, is up to us. It is, however, a meaningful factor in our development, our success, and our happiness.

Good News . . . and More Good News

We have already discussed Wordsworth's reaction to the Industrial Revolution. It is safe to say that technology has advanced with amazing rapidity to produce systems, ones that are increasingly complex and have multiplied by a factor of ten the number of problems common to every human being. In NATI terminology, we speak of Polarity, or opposition, as the psychological engine that drives human problems. As formidable as these factors can be, we can change our lives—our very paradigms—by our thoughts (or more specifically, by our point and frame of reference). Using thirteen innate intelligences possessed by every human being, we can learn to live in inner harmony using nature as our model. *By the identification of absolute principles, we then have a basis for proceeding to any issue at hand, as well as for our development!*

The Biology of Understanding

As mentioned in the introduction, I have studied physics, math, biology, philosophy, theology, and sacred sites around the world. Of all the epiphanies and revelations I have had, none has ever rivaled the one stemming from the research shared with me by Dr. Derald Langham, who identified the geometric structure of the development of the cell. As a cell develops, there are three parts to its pulsing motion during its Creative Phase. There are six parts to the cell's Organizational Phase, exhibited by wave motion. Finally, the Functional Phase, characterized by spiral motions, displays four parts. The thirteen parts of cell development correlate with the intelligences described by Natural Thinking & Intelligence. The direct correlation between this geometric model and man's thirteen intelligences shows that man's thinking and intelligence are not only derived from, but are a part of, nature itself. This is why "nature as a life model" is proposed.

Exciting things can happen once we shift our paradigms and accept the fact that it *is* possible to make our personal voyages through life according to the natural laws of the universe. Therefore, we can minimize the factor of problems by utilizing the potential energy always present. This is achieved by converting problems into healthy goals and challenges.

Sound difficult? It's not!

Nature is configured to accommodate the achievement of potential and the evolution of higher consciousness. Nature is accessible to everyone, an idea that Wordsworth spent a lifetime trying to communicate in his poetry. With the use of our thirteen intelligences and the principles of nature described by NATI, we now have a coherent system to reconnect with nature regardless of the complexity of technology or the frenetic pace of modern life.

It's why we're all here.

Theory of Living Systems

So where does all the discussion concerning Nature as a model for Understanding lead us? The answer is to General Systems Theory, a model of reality utilizing biological and physical sciences as basis for human behavior and thinking. We will discuss this later more fully.

Suffice it to say here that this theory of living systems is an all inclusive, comprehensive model of life. It distinguishes itself from anything else by numerous factors. Here are just a few:

- With living systems, things are not viewed as separate.
- Everything is a part of a whole, a greater whole.
- It generates a web of relationships between everything.
- It demonstrates the essences of things.
- It has emergent properties as we discussed in the introduction.
- These living systems have an inherent order.
- They represent an outline for the way we lead our lives.

What NATI brings to the mix is an identification of Nature's principles as well as a universal language for understanding and utilizing those same principles. Moreover, NATI 's philosophical basis of Potential and its development as a life focus is likewise key!

Summary of Nature

For Aristotle, the world was one of self-development. He identified the principle of growth and change as "an essence that contains self movement."[7] This inner essence tends to cause people to actualize their potential and become whatever it is in their nature to be. This is what decoding potential does—identifies what that essence is within each of us! This is best accomplished by identifying one's point of reference.

In 1808, John Dalton's *New System of Chemical Philosophy* provided us with an open-ended resource for a framework of concepts, expressions, and structure that are utilized as the basis of subsequent matters.[8] As you will read, this is what the NATI structure does. Nature and its principles, identified by Decoding Potential, stand above and apart from human law.

Nature makes humankind's relationship with its creator clearer, less restrictive, and a great deal more practical. Current systems attempt to understand God, themselves, the universe,

and humankind through subjective and limited notions and laws.
Decoding Potential through nature's principles utilizes objective, universal concepts and frees the individual from restricted,
limited notions.

Chapter Two

Introduction to Natural Thinking & Intelligence

What Is Natural Thinking & Intelligence?

Natural Thinking & Intelligence is the identification of the basic principles of nature and reality. But what exactly does that mean? It is, after all, a pretty sweeping statement to make.

Imagine if you found the secret structure of human consciousness and behavior—the common elements within each one of us. Next, imagine that this structure enabled us to understand *how* these common elements worked. This would be quite a find, for we could overcome all previous failures and accomplish what we could previously only dream of. We would have not only a way, but the only *true* way, of discovering our purpose and direction in life, whether it be physical, mental, emotional, or spiritual.

For the sake of simplicity, we are going to collectively label these basic principles as NATI, for Natural Thinking & Intelligence. The logical question, therefore, is exactly what *are* these principles that define our structure and reflect the nature of reality?

There are thirteen of them, each belonging to one of three different categories, labeled Creative, Organizational, and Functional. Furthermore, all of these basic principles have important characteristics.

- They are neutral and have no frame of reference. They are not aligned with any position or opinion.

- The identification of a principle (or a combination of them) can define an individual's traits, subjectively or objectively.

- They are invariant. That is, they are absolute and do not change. Only the perception of them changes.

- They cannot effectively be replaced with anything else.

- They are universal. They are everywhere we look and form part of every human conscience and behavioral pattern one can identify.

- They are part of a closed system. This means they are an entire system within themselves and therefore closed to outside influence. They have the ability, however, to operate as open systems with infinite possibilities.

NATI Principles as an Intelligence

In this chapter, we will identify the principles mentioned above and their various interpretations. We will also show NATI principles as intelligences, innate parts of human consciousness and behavior.

In subsequent chapters, we will also see the following:

- A structural and philosophical discussion of how NATI is found in the concept of potential as it relates to the universe and humankind.

- An examination of Polarity (or opposites) in nature and science as seen in various concepts, such as Einstein's Theory of Relativity, as well as in Eastern philosophies.
- Examples from real life that demonstrate NATI principles as the basis of everyday occurrences.
- NATI paradigms and formats generated from NATI's principles and its basic operating systems:
  - The Human Character Formula (awareness + belief = the character of our expressions, or a + b = c)
  - Our Current Operating Procedure (COP)
  - Our Point of Reference (POR)
  - Our Frame of Reference (FOR)
  - The process by which we develop our personal identity though preferences, performance, Personality Orientations, Core Human Dynamics, and what NATI calls the Great Restrictors.
- What NATI can actually do for you and how to identify your personality profile (your strengths and weaknesses in relation to this natural system of thinking and intelligence).
- How to find and implement your resources and generate others you never believed you could possess.
- A look at the concept of development as a universal objective not only for individuals, but also for the entire species—the master plan of the Creator, if you will.
- The notion of wholeness as an ultimate intelligence and the importance of identifying and integrating all forms of information into our daily lives.

Defining Intelligence NATI Style

If NATI is a system that uses thirteen kinds of intelligence, just exactly what *is* intelligence to begin with? Let's start with some basic notions. Simply stated, intelligence is the ability to gather, recognize, and integrate data into wholes and then apply the data for the purposes of development. The greater a person's ability to do these things, the more complete and successful he or she will be. What we learn about intelligence as we grow up, however, is riddled with stereotypes, and like most stereotypes, they are incorrect. We automatically assume, for example, that the nerd carrying a stack of books is intelligent, while the campus jock is inherently dumb. Neither of these assumptions, of course, is necessarily true. With Natural Thinking & Intelligence, there are categories, devoid of stereotypes, that can adequately describe every form of intelligence that, until now, has been either overlooked or categorized as something *other* than intelligence.

Despite much study, scientists and psychologists have yet to settle on a precise characterization of intelligence. Regrettably, this hasn't dampened enthusiasm for the design and application of standardized tests, such as IQ tests, resulting in the kind of shallow definitions of intelligence found in school systems. Given the narrow limits we place on understanding intelligence, it is helpful to consult an accepted definition of the term, one found in the *Oxford Companion to Philosophy*.

> A family of intellectual traits, virtues, and abilities occurring in varying degrees and concentrations. An intelligent creature is one capable of coming up with the unexpected. An intelligent person is one in whom memory and the capacity to grasp relations and to solve problems with speed and originality are especially pronounced.[1]

As this definition implies, intelligence is not limited to the ability to store information, nor is it simply a matter of attaining verbal, social, or mathematical skills. More accurately, intelligence is *the ability to recognize data and apply knowledge.*

Let's look at intelligence, therefore, in terms of recognition and application. The recognition of data, of course, implies that data is ultimately stored, but a person's level of intelligence cannot simply be measured by how much information is stored in someone's brain at any given moment; rather, the level of intelligence is determined by the mind's ability to *apply* what is known to any given situation. Formal education attempts to address this by teaching critical thinking skills, but while critical thinking is mentioned in virtually every curriculum statement in the country, such skills are rarely practiced. To administrators and school boards, it is more important to finish the material outlined in lesson plans than to help students apply what they have learned in any meaningful way. Intelligence is largely approached in quantitative rather than qualitative terms.

Application, therefore, is crucial. Ultimately, intelligence is *the act of recognizing information and then placing it into a meaningful context.* To a greater or lesser degree, we all do this every day. From child to adult, from Pygmy tribe member to corporate executive, everyone expresses qualities of Natural Thinking & Intelligence.

If information is not correctly assimilated in the acquisition stage, however, the chances increase that raw data will be stored in our brains incorrectly. The result is that our attempts to apply what we learn produce confusion rather than enlightenment. That's why it is necessary to understand the stages involved in what might be termed the evolution of raw data. In Natural Thinking & Intelligence, the best way to follow the process by which raw data is transformed into genuine understanding is through RIKU, (short for Raw Data, Information Knowledge, and Understanding). The four steps of RIKU are as follows:

1. Raw data is experienced as isolated, disconnected bits of information.
2. Information begins to emerge as the bits of raw data form recognizable, meaningful patterns.
3. Previous knowledge then recognizes the connected information.
4. Finally, understanding empowers us to know how to apply the recognized information.

RIKU is a straightforward way of understanding how data moves from recognition to application as noted in the definition from the *Oxford Companion to Philosophy*.

This transformation of information into understanding is no different than the process of creating music, in which raw data consists of notes on a musical scale. By placing these notes together in various combinations, specific tones are constructed. When enough tones are juxtaposed, the result is what we call music. The same holds true for the alphabet. Individual letters form words, then sentences, and, eventually, entire books. In both examples, isolated bits of information are first recognized and then applied.

Information is sometimes incorporated into our understanding in a rather shallow fashion, while at other times we process data in far greater detail. Hearing background music in an elevator is a completely different experience than consciously listening to the complex melodies and repeated motifs of a symphony. As you learn more about Natural Thinking & Intelligence, it will become clear just how important it is to have a full understanding of any piece of information in order to successfully apply it to a larger context, regardless of how inconsequential the initial data may seem.

Considering the amount of data we all must process each day and the number of decisions we must subsequently make, the potential for improving our lives is truly limitless if we use a system of thinking modeled after nature itself.

This notion of RIKU is also apparent in the structure of this book, which first presents raw data that eventually leads to understanding.

The Three NATI Groups of Intelligences

We have already mentioned that the thirteen NATI intelligences fall into three categories: the Creative, the Organizational, and the Functional. Together with Polarity (or opposition), everything that exists falls within the scope of these groups. As we shall soon see, potential has the ability to create, organize, and function, and it does so because it includes the idea of opposites in its dynamic structure.

Creativity implies that we are focused on something, have a belief about it, and then express it in one way or another. The Creative Group will show how three intelligences describe these activities.

Accordingly, six intelligences in the Organizational Group will demonstrate various activities—all organizational in nature—such as synthesizing parts into a whole, prioritizing, measuring, making assessments, understanding and using feedback, finding patterns, and establishing procedures. This is by no means a full range of organizational activities, as we shall see in the very next chapter.

Finally, the Functional Intelligence Group demonstrates the only four ways in which the human mind can function: physically, mentally, emotionally, and intuitively.

As we move forward, remember that these three groups correspond to the natural model of cell development and are therefore parts of the very fabric of the universe.

It should be noted here that there are two segments of our NATI construct. One is the discovery and identification of our Natural Intelligences. The second part is Natural Thinking Systems. While you do need to understand the intelligences in order to utilize the thinking aspect, *it is not necessary that you understand the thinking aspect in order to utilize the intelligences on their own.*

Chapter Three

The Thirteen Natural Intelligences

The Creative Intelligence Group

There are thirteen intelligences found in the Creative, Organizational, and Functional categories. The first three intelligences belong to the Creative group.

The Focus/Awareness Creative Intelligence

Focus, or Awareness, is one of the most significant intelligences in that it is, for all intents and purposes, where everything begins. It is the onset of consciousness and is connected with all other types of intelligence since everything we do requires some level of focus or awareness. Athletes, for example, use an incredible amount of focus to enter what they call a "performance zone," a mindset in which they are able to block out all distractions and engage in peak performance.

One indication of the importance and power of Focus is the

body's response to imminent danger. Numerous studies have demonstrated that humans sometimes perceive events in slow motion during the final seconds before car wrecks, airplane crashes, or any number of accidents that pose a real threat to survival. Although the precise biological mechanisms responsible for this phenomenon are not completely understood, some researchers theorize that the brain becomes hyper-focused, literally kicking itself into a different gear so as to access the greatest amount of information in the shortest time available. Focus is an evolutionary mechanism that aids us in a variety of ways, the least of which is survival.

Focusing is concentration, while awareness is consciousness. Although they are two different factors, they both fall under the same principles that Pythagoras described as the power of will! For the sake of clarity and simplicity, NATI uses the terms Focus/Awareness.

In the early 1960's, Prof. Eugene Gendlin of the University of Chicago was determined to find out why some therapies were successful and others were not. The result was that the successful clients were actively engaging in the process. They were utilizing more than just logic, or as he put it, "They didn't just stay in their heads."[1]

According to Ann Weiser Cornell, Ph.D, in her book *The Power of Focusing*, focusing is a natural skill that was discovered, not invented. "It is a very broad purpose skill."[2]

Whether we decide to write the great American novel, paint a picture, invest in the stock market, or make out a grocery list, every activity we engage in calls for some degree of awareness

It is imperative the reader understands that the Focus principle is one's Point of Reference, or POR. This is the starting point from which all events, behavior, and understanding emanate.

The Beliefs/Concepts/Perceptions Creative Intelligence

The next kind of intelligence in the Creative Group is Beliefs, Concepts, or Perceptions. (Note that in many subsequent sections, a particular intelligence may be referred to by more than one name to better describe the continuum of concepts covered by the intelligence. It will nevertheless represent only one of the unchanging thirteen intelligences, regardless of its synonyms.) The intelligence of Concepts refers to how we interpret reality. Everything we are confronted with in life requires that we adopt some concept about it. We form beliefs and perceptions about the news we hear, the people we meet, the faith we practice, the government leaders we elect—about everything, in fact. Our brains are wired to evaluate the information we encounter.

It is important to distinguish beliefs from ordinary perceptions, however, since perceptions may be temporary or lack the intensity of an actual belief or concept. We may form an unfavorable opinion about a new co-worker—a perception—that rapidly changes when we get to know that person. This stands in stark contrast to the way people cling to belief systems. The old adage of "never argue about religion or politics" is based on the tenacity with which people cling to a given concept. The history of warfare is largely a chronicle of clashes between political or religious belief systems.

Education is another example of a belief system that elicits an almost visceral response in many people. We approach the present-day curriculum much as we did at the beginning of the twentieth century. The philosophy behind learning then, as now, was that it was possible to know everything that was knowable. English, mathematics, history, foreign languages, and literature were the main disciplines studied. Science and philosophy were peripheral and were mostly taught to university students. Despite an explosion of information and technology in the last several decades, this core curriculum has been updated only slightly. Educational philosophy is still based on the concept of a core curriculum, now over a century old, which purportedly prepares students for any experience or job in the world. Innovative thinkers who challenge the idea of a core curriculum make little headway in establishing newer, more relevant curricula.

The intelligence of Beliefs or Concepts, however, is not limited to "isms" or doctrines. Again, this creative intelligence refers to how we interpret every aspect of reality. We can have beliefs about patriotism, reality television, or household cleansers. In short, Beliefs is a vital component of human intelligence, one that drives creativity at a deeply fundamental level. *Most importantly, it frames reality, as well as our image of reality. It is our Frame of Reference, or FOR.* Later on, the reader will see the profound impact that Beliefs and Concepts have on daily reality. Focus and belief are two of the most powerful principles of understanding!

The Communication/Expression/Creative Intelligence

The final Creative Intelligence is Communication, or Expression. In simplest terms, this intelligence refers to how we communicate. There can be no doubt that this is an elemental, innate form of intelligence, for there is little in life that does not involve the exchange of information. Information is exchanged among atoms, plants, animals, human beings, and the systems and institutions humans create.

Many historical figures are known primarily for the ease or wit with which they communicated certain ideas. Consider Mark Twain, Will Rogers, Teddy Roosevelt, Ronald Reagan, or Billy Graham. They made their marks on history because of their insightful, humorous, or colorful rhetorical styles. Sinister figures of history are also known for their charismatic or mesmerizing manner of communication. Dictatorships depend on propaganda, and propaganda is nothing more than the skillful manipulation of information and the matter in which it is communicated. Adolph Hitler could not have built the Third Reich without the ability to mesmerize millions of people with his words and body language.

Call to mind almost any revered historical figure and you will see someone who was able to advance a belief or philosophy because of the ability to communicate effectively in one medium or another.

Another indication of the importance of this intelligence is that we are often impressed not by *what* is said, but how people say it. Each and every day, we both send and receive thousands of communications to the people we interact with, from friends and family to business associates. A simple transaction with a store clerk can run smoothly or become an exercise in anger depending on the communication skills of both buyer and seller. How often have arguments ensued because one of the parties was perceived to speak with aggravation, condescension, or sarcasm? By the same token, people harboring widely divergent opinions can communicate quite effectively when they are skillful at expressing themselves and are able to do so without unnecessary emotion or judgment. Again, what is important is not always the actual content of communication as much as the manner with which it is conveyed. Consider the con artist. His data may be untrue or distorted, but his delivery is compelling.

This holds true for other forms of communication as well, such as body language. Various postures and poses can convey great sympathy or openness, while others—a rolling of the eyes, a frown, or a simple turning away from a speaker—can convey as much invective or rejection as the spoken word. Think about a parent's penetrating stare at a misbehaving child.

Information is also exchanged in the systems and institutions man creates. Systems of education, government, and business could not function without a constant flow of information. People receive information on the stock market and decide whether or not to invest in a particular company. The number of overall investments made is data that regulates our economy and gives people further information on whether or not to buy or sell certain stocks. The more accurate the data expressed to the investor, the greater the successes.

The same type of thriving, open interactions can be seen in many other systems. Students, teachers, parents, and administrators communicate with each other in numerous ways that constitute the educational system. Likewise, elections, the passage or repeal of laws, public debate, and the judiciary system all entail communication that results in the institution we call government.

Another vital point here is that the communication intelligence is further demonstration of one's expression of his references.

The Human Character Formula

Let's look at a key factor in the systems application of the three Creative Intelligences. Communication ultimately goes beyond the manner of its delivery, for expression has its basis in the science and mathematics of geometry. *Thus we will now see how what you focus on and what you believe about that focus will always equate with how you express it.* This relationship among the Creative Intelligences can be expressed in what NATI terms the Human Character Formula of A + B = C, or

A (awareness) + B (beliefs) = C (character of communication)

In my personal studies, I discovered repeatedly that this single formula resonated

perfectly with key ideas from many different disciplines. The character of our communications, for example, can be traced to the ancient mathematical formulae known as the Pythagorean Theorem (the sum of the squares of two sides of a right triangle equals the square of the hypotenuse). Pythagoras was not only a mathematician, however, but also a thinker and a philosopher. While his theorem expressing the relationship between the sides of a triangle is still widely taught, Pythagoras himself correlated the theorem to his own philosophical thesis of

will + belief = expression

Since a person's will is an outgrowth of his awareness (and requires focus and concentration), we see how elegant and mathematically precise the Human Character Formula is. This (and thousands of applications over twenty-five years) is further incontrovertible evidence that NATI principles are part of nature itself.

The three factors in this formula are also mirrored in many other fields of study, such as psychology, philosophy, Eastern religions, and others. While many of these concepts are a bit weighty and would require pages of explanation, we can see a sample correlation in the following example from psychology. The three contributing factors are nature, nurture, and creativity. Nature corresponds to unconscious instincts (basic awareness), nurture corresponds to past conditioning (the elemental beliefs we acquire), and creativity corresponds to the drive we make toward potential (expression or communication). (The interested reader is directed to Jung, Maslow, and many others.) An example of this would be as follows: When I think about (focus on) my kids, I want to make them secure (a concept), so I work harder (expression).

It is worth noting that the Bhagavad-Gita describes three primal components of life that also correspond to nature, nurture, and creativity. As with all NATI principles and intelligences, the character formula is a code written into the fabric of the universe. The commonality of all open, dynamic systems is that they describe a visceral, natural drive toward creativity and expression. This drive is ultimately a movement toward the development of potential and is intrinsic to all of us. Communication is the Expressed References (ER), or the ones we eventually manifest in our behavior.

"I Need to Get Organized"—A Common Issue in Today's Society

The next six intelligences comprise the Organizational Group. They are a consolidated group of elements that culminate in the idea of structure. We utilize any of these elements whenever we undertake an organizational task.

The Laws/Models Organizational Intelligence

The first organizational intelligence is Laws or Models. This intelligence refers to the adoption of an image or picture of what works. While society is often indifferent to modeling, NATI prefers to say that there is a certain amount of truth in the cliché that "imitation is the sincerest form of flattery." Indeed, modeling is severely underrated as a way of improving human functioning. There are literally hundreds of how-to books about writing that have been published, and virtually all of these manuals spurn imitation as a way of developing a unique narrative voice and style. This advice to shun imitation betrays an ignorance about the creative process, for it is precisely through imitation of several styles that a writer can synthesize a new voice, taking a little from each author studied or imitated until various stylistic elements add up to a new kind of prose that is totally distinctive. This kind of modeling actually goes back to the Greeks, who believed that the imitation of models was the best way to attain perfection.

Business is another area where the use of models is critical for efficient functioning. Businesses utilize models for numerous applications, such as methods of organization. In terms of management, for example, most businesses are modeled on concepts of authority rooted in a chain of command. Businesses also rely on economic models of free enterprise, profit, and supply and demand. Additionally, many corporations have research and development sections that produce new products based on given rules, laws, or standards set forth by the marketplace.

Quite simply, Models show us what works. While it may sound simplistic to think of Laws or Models as a form of intelligence, consider something as rudimentary as making a decision to eat a balanced diet. People who are overweight cannot shed pounds without a safe diet plan supplemented with exercise. While fad diets and weight loss pills flood the market daily, the most proven method is to cut back on saturated fats while exercising on a regular basis in order to burn excess calories. Effective models of dieting show us what works and what doesn't in our efforts to maintain health. We can choose a fad diet or unsafe diet pills, but the model is up to us. However, the best model will be the one that functions at the highest level, namely, one that maintains the basic factors of safety, effectiveness, and efficiency.

Even the process of biological development seems to hinge on the concept of Models. The process of natural selection increases the chances of development for some members within a species that has already experienced past success in adapting to its genetic environment. The very mechanism of evolution is based on the concept of using what has already been proven to work. *Modeling represents the very foundation of natural selection.*

In short, nature's own economy and wisdom, aimed at potential growth, seems to place a very high premium on the intelligence of Models. When we organize after adopting a plan of action, the first step is to establish rules, laws, or a model to govern the plan. Modeling represents the parameters of our references.

The Detail Organizational Intelligence

The next organizational intelligence is Details. This is a straight-forward but essential intelligence because it has pronounced implications for the synthesis of the thirteen intelligences into a system in which all of human life functions. The idea of Details relates to the individual's ability to perceive specifics, levels, or separate items in the context of a larger system. It is related to the breakdown of an object or idea into its constituent parts.

The importance of Details is evident in everything around us, and yet we probably don't recognize this intelligence, a clear case of not seeing the forest for the trees (or more aptly, the whole for the parts). Virtually everything in existence is comprised of components, although most of us generally focus on "the whole" rather than its constituent parts. The same is true for issues or ideas. We focus primarily on a general concept rather than the details (or nuances) that may underlie a particular concept. Most of us would readily agree that the Constitution is a brilliant document that validates the great American experiment in democracy. A majority of people, however, would not be able to name or discuss its amendments or the Bill of Rights.

While most people should be better versed in the foundation of their government, it must be conceded that a total assimilation of Details is impossible in many, if not most, areas of life. It's only natural that we focus on the whole, for we would never get through the day if we stopped to examine every part or detail we encountered. All matter, for example, is made of atoms, which themselves are broken down into smaller particles, such as protons, neutrons, and electrons. As creatures in the macro-world, however, we do not normally focus on the chemical or scientific nature of reality. We are more than content to deal with three-dimensional objects such as tables, lamps, and chairs.

The devil is in the details, as they say. The communication of information presents another excellent example of Details. Books are made of pages, upon which are printed words formed from individual letters. The valuation of a home is another example. We arrive at a final value by looking at given details: its condition, location, size, property, taxes, nearby schools, and other factors. The fewer details one considers, the greater the chance for a mistake in value!

People who understand the importance of Details can relate to an issue in greater detail. People are considered erudite when they go into specifics. Individuals who think or act in generalities don't get very far in convincing others or making a point, nor do they understand a subject as well as they could if they pursued details!

Details are therefore important in the humanities, arts, and the social sciences. Philosophers and psychologists break down various ideas and concepts into steps or keys. Playwrights divide their works into acts, while composers divide symphonies into movements. Details are just as important in thinking and aesthetics as they are in science and technology. Indeed, in the mind science that is NATI, we have thirteen separate intelligences—thirteen parts—one of which we are discussing at this very moment.

We shall see in later chapters that certain aspects of Systems Theory demand that we look at how components are synthesized, but without the existence of specifics to begin with, any discussion of wholeness would be meaningless. For this reason, Details is one of the thirteen fundamental intelligences of the universe. (My research shows that women tend to be much better at details than men. My wife Angela is much better at details than I, and she reminds me of that—with details—frequently!)

A final word here: note that this aspect represents the details of one's references.

The Order/Processes Organizational Intelligence

The next NATI intelligence is Order, or Processes. In its simplest terms, this intelligence encompasses procedures and all patterns of action. It is literally how we do things—the procedures we follow or the patterns we utilize to accomplish tasks both large and small.

Daily routines are examples of processes we engage in to accomplish various tasks. Every morning, many of us get up and go through certain rituals—grooming, dressing, and eating breakfast—so that we'll look presentable in the workplace. We then drive to work or use mass transportation that runs according to a certain schedule so that we will arrive at work on time. Throughout the day, we then adhere to more schedules in order to accomplish a quantifiable amount of labor. This is done for the dual purpose of furnishing society with goods and services, as well as providing a personal income, making possible, among other things, the continued practice of the initial processes of getting up, grooming, and going to work. This is order!

It is not surprising that many people, if not most, complain from time to time about the predictability of their lives that stem from various procedures. The parent laments the seemingly never-ending routine of cleaning house, driving children to school, preparing meals, and then jumping into a minivan to pick up the kids and bring them to afternoon soccer practice. Similarly, it is not uncommon to hear a wage-earner express frustration at being caught in the "rat race," a cycle that all too often seems to exist only for the purpose of perpetuating itself. Certain repetitive procedures can literally cause depression in some people or the more common complaint of "Boring!"

By contrast, NATI approaches procedures with a mindset that works in harmony with nature, enabling people to transform frustrations over routines into action and achievement. A quintessential example here is the way in which the Japanese have integrated holistic health into their workplaces and factories. Japanese corporations schedule frequent breaks during the workday for rest and exercise. Many companies even permit time for recreational activities or hobbies such as bonsai, which is a novel approach inasmuch as bonsai itself requires focus, concentration, and order, and yet it is perceived to be relaxing. This more personal approach to employee-management relations may seem strange to the Western corporate mindset, but the reality is that the Japanese are using various NATI intelligences to enhance productivity by enabling the worker to realize greater potential. The Japanese model indicates that order need not constitute an imposition on our personal growth.

In the NATI universe, therefore, it is important that we understand that there exists a natural order, one of the key components of which is given, absolute procedures. When we are able to grasp the inherent order in a system, we are often able to predict what is coming next. In climates where the four seasons are discernible and show the classic variations in temperature and precipitation, it is easy to know what weather pattern is coming next. Knowing this pattern enables people to plan their lives in many ways. In business, department stores know when to put season-appropriate lines of clothing on display. In healthcare, professionals know when to prepare for flu season or the treatment of sunburn. These are all very positive aspects of order, process, and predictability.

Procedures can describe incredibly complex degrees of order as well. Although patterns unquestionably apply to the mundane—the manipulation of everyday objects or the recognizable sequences we use to accomplish various tasks—they are equally valuable in understanding scientific, psychological, and philosophical ideas. If we wish to study any branch of math or science, we must go about it systematically, allotting time to study, assessing the amount of material to be covered in each session and working through problems to test our understanding of concepts. Indeed, teachers and students who are successful use various procedures at every stage of learning.

In short, the intelligence of Order or Processes shows us the steps involved on the paths we travel. The better ordered we are, the more efficiently and effectively we function. In my studies, I have discovered that men are much better at Process than women. I never need to go looking for my keys, while Angela . . . well, she is detailed!

Note that this intelligence also relates to one's pattern (or patterns) of reference. This pattern emanates from our various points of reference.

The Measure/Organizational Intelligence

The fourth Organizational Intelligence is Measure. It is analogous to prioritizing issues, events, or the daily actions we engage in. It pertains to the significance or depth of our understanding. In certain respects it relates to judgment. We see this intelligence at work all around us, both at work and play. Whenever investors seek to purchase something, they judge the risk and return they could realize. Some may throw caution to the wind, while others may seek numerous exit strategies. Depending upon the individual's success or failure, priorities may be changed. The priority may be rearranged, the amount of investment may be altered, or the type of investment may be modified.

Politics also exemplifies Measure and Assessment. A responsible elected official should always ask what impact a proposed law will have on all his constituents. Is the law capable of enforcement? Will it unfairly affect some while giving others advantage? Does its rationale serve the common good? Not all laws, of course, are deemed just. Any judicial interpretation of laws after they are passed constitutes a very formalized assessment. Indeed, without the intelligence of Assessment, there could be no system of checks and balances whereby each branch of American government modifies the performance of the other two branches.

Prioritization is another way of expressing this kind of intelligence. A general might feel that it is appropriate to deploy troops to one theater of operations before another. In government, it might be necessary to assess methods of taxation before enacting laws on an action such as Medicare. In everyday life, we must all select which appointments, errands, or chores take precedence over others. Without these kinds of assessment, nothing would ever be accomplished.

In the long run, all assessments must be viewed as relative. In fact, relativity is an important aspect of Measure or Assessment. Interestingly enough, Einstein's Special Theory of Relativity keeps only the speed of light as a constant, while measurements of time and mass vary according to an observer's position relative to light. Therefore, relativity is a key factor in the establishment and implementation of one's point and frame of reference. In NATI, we therefore say that relativity is subject to the law of measurement. Because relativity contains the concept of zero and infinity (and everything in-between), all aspects of measurement must be included in our relativistic universe. One thing is contingent upon another. The irony here is that while measurement itself is relative, open, and flexible, the principle of measurement is closed and absolute. It is always present! We will read later how these open and closed systems function.

In the all-encompassing system that is NATI, we demonstrate to people how judgment and priority bring structure while reducing chaos in their lives. An example here is training oneself to adhere to our principles in everyday reality.

Measure is critical if we are to transform what Taoists call the living void, filled with infinite potential, into recognizable patterns of reality. The better one's ability to measure, assess, or prioritize, the more accurate his discernment. *One of the hardest things to do in organizing is to stick to priorities.* In my experience, people change priorities faster than they flick a light switch. Flexibility is fine. Chaos is not!

Note that measure, assessment, and priority relate to prioritizing one's references.

The Feedback Organizational Intelligence

The fifth intelligence in the NATI matrix is Reflection, Mirroring, or Feedback. This is an extremely visceral intelligence. It is derived from the age-old oriental notion that whatever bothers or disturbs us to any significant degree is really some internal factor being reflected back to us. In other words, whatever unsettles us concerning another is actually something within ourselves that is wrong or incorrect and needs attention.

A good marital therapist will always help patients explore the possibility that complaints about their partners may originate because of some inadequacy inside of themselves, not their husbands or wives. A basic tenet of psychology is that people have a tendency to project their own unhealthy behaviors onto others, in whom they then see the behaviors mirrored, albeit they are completely unaware of this process.

Husbands and wives often claim that they do not receive enough attention (or the right *kind* of attention) from their partners. If a husband or wife feels neglected, he or she may well be neglecting something important in his or her spouse. A husband may not be getting enough credit for the time he spends at work, while a wife may not be receiving enough praise for her role as mother, homemaker, or provider of a second income.

Two excellent examples of the phenomenon of Mirroring involve my good friend and colleague, Terry Anderson, who was the AP Bureau Chief for the Middle East. Terry was held hostage in Beirut from 1985 to 1991 by Islamic militants demanding the release of seventeen Shi'as convicted of bombing the French and American embassies in Kuwait. In his book *Den of Lions*, Terry writes of the tension that existed between himself and another hostage.[3] After Terry precipitated a confrontation between the two, the other hostage left the cell to wash up. In the absence of the alleged troublemaker, Terry urged his fellow hostages to do something since his nemesis was, in his opinion, being obstinate and inflexible. Much to his surprise, the other hostages told Terry that he himself was the one being obstinate and inflexible and that he was as much responsible for the tension as the other hostage, if not more so. In his book, Terry compares his surroundings while a captive to "a house of mirrors."

Many people immediately challenge or dismiss this intelligence because it hits too close to home. Even Terry Anderson rejected the idea of Reflection and Mirroring at first. It wasn't until I reread his book and discussed it with him that he began to give it credibility.

The same thing happened when Terry and I conducted a seminar in Rye, New York in 1998. After explaining the element of Mirroring to the group, I was instantly challenged by several people who attempted to refute the notion. One woman in particular adamantly rejected the entire thesis. I then did what I always do in a situation like this: I looked into the mirror. I asked the woman to give me an example of something that really bothered her, preferably something recent. She related a situation concerning a new job she had just begun, one in which several of her fellow staff members were treating her with arrogance and disrespect—or so she said. Within ten minutes she openly admitted that the very first time she walked through the door of her new place of employment, she had conducted herself in an arrogant and condescending manner!

To reiterate, this is a very visceral science, and I have never seen it fail to provide a valid insight. As to why it exists or what metaphysical meaning it holds, I have my opinions. It may be that God, or a God-force, wove it into the design of the universe for the purposes of development. Development is a universal principle and should be our ultimate focus because consciously or unconsciously, we are all striving for it, and Mirroring exists to show us the way. By using this intelligence, we can become better human beings, more sensitive and aware of the things we do in our lives. Granted, it is a difficult concept to deal with, for nobody is perfect—only the path of development is. People who reject this notion of mirroring usually don't accept accountability very well because it exposes them to their inner selves, and this frightens them. It is a vital organizational intelligence in that it gives one immediate feedback as to what's going on, what's wrong, and what action to take.

Note that mirroring relates to the feedback of one's references.

The Wholeness/Synthesis Organizational Intelligence

The sixth and final Organizational Intelligence is Wholeness, Integration, or Synthesis. This is the ability to synthesize information and place it into a larger picture or system. It is the ability to perceive unity and integrate key components into "the big picture." The world is positively replete with examples of wholeness. One who is gifted with any degree of wholeness is truly blessed. The ability to integrate information or events into a meaningful picture is actually the highest degree of understanding. Some people may equate this intelligence with liberalness because of its inclusive nature. This would be inaccurate, however, and betrays a complete lack of understanding. Where integration occurs, opposites (Polarity) obviously come together. They then compliment each other. This is representative of the person who knows how to take negatives and turn them into positives.

Let's look at this element through a real life situation involving our investment company. My son Rob found a property whose owner wanted to sell at the highest possible price as soon as possible. To achieve this, however, the property needed to go through extensive approvals that would take a good deal of time. However, he needed money right away to buy elsewhere. We knew the minimum value of the property and were willing to wait to maximize the value. We also wanted to insure liquidity in case something else came along. Resolution? We signed a long-term contract with sliding price levels dependent upon the degree of approvals. In the contract we provided the seller with a mortgage on the property for sixty-six percent of the purchase price at the prevailing rate. This insured our investment and our liquidity while at the same time providing us with a good rate of return on our invested capital. A lot of negatives issues were integrated and therefore resolved.

Insofar as human nature is concerned, synthesis and wholeness relate to integration. Being accepted into a group represents wholeness. When we become part of an athletic team, a company, a marriage, or a community of some type, we are experiencing the process of synthesis.

Given this book's thesis of nature as our best model, the final vision we shall attain of all thirteen intelligences as part of the NATI matrix is perhaps the best example of synthesis and integration. By now, the reader should be able to see the interconnectedness of all intelligences. We are aware of details or concepts, for example. We may model them, prioritize them, and ultimately place them into the context of something larger about which we may or may not have strong beliefs, causing us to take some kind of action. If a man building a house chooses a type of wood, he will do all of the above. He will decide how much wood to use, measure it, determine how to use it according to a blueprint, nail it into place, and decide if he approves of the way it looks.

Note that this principle relates to the integration of references.

The following summarizes what we have said thus far regarding the Organizational Group.

ORGANIZATIONAL

- Formative
- Structure

MODELS	ORDER	MEASURE	MIRROR	WHOLE	DETAILS
Laws		Levels		Unity	
Form		Degrees		Entirety	
Rules	System	Relativity		Overall	Components
Identity	Format	Emphasis	Reflection	Collective	Individual
Support	Pattern	Intensity	Complimentary	Group	Segments
Basics	Discipline	Significance	Response	Composite	Separate
Agenda	Process	Comparison	Interaction	Continuous	Unit
Models	Procedure	Judgment	Resonance	Generative	Sector
Principles	Steps	Probability	Feedback	Connected	Segregate
Parameters		Infinite		Integrate	Subdivision
		Dimensions		Synthesize	Parts
		Assessment		Total	
		Priority			

WHOLE/ TOTAL
The highest level of understanding. The more types and styles you can embrace, the better you are. The greatest intelligence is when every part fits into an issue. One can see the entirety. Everything is recognized for what it is. It is the viewing of the entire picture with all one's parts.

MIRROR/ FEEDBACK
We are constantly surrounded by reflections of ourselves, especially the negative. Feedback is a reactionary method of knowing and learning. If it is experiential (vs. insightful/qualitative, etc.), it may well miss long-term knowing. Mirroring exposes the problems with the path to an objective.

PATTERNS/ ORDER
These are processes, cycles, steps, etc. in any structure, including chaos. This is the place of paradigm shifts, the path one chooses. At some point in a process, underlying concepts clash and hidden patterns emerge. Patterns are mostly rote. They should be analytical. Information (intelligence) changes patterns, which are the steps towards an objective.

MEASURES
This is judgments, priorities. They shift and ebb according to our focus and beliefs (ethics). Relativity is perhaps the greatest, most valid measure. By matching functional and creative principles with core dynamics, a true sense of priorities emerges. Also, some interesting concepts emerge.

DETAILS
These are the parts or units that one implicates with an issue, or the segments comprising the totality. They are the bytes of information necessary to accomplish an objective or to attain a goal.

MODELS/ RULES
Laws or models are at issue here, indicating what rules or structures are in place or to be followed. Imitation is another interpretation here. Models are examples of what can or should be achieved or followed. These are the standards for adherence to a focus or concept or an objective/action.

The Functional Intelligence Group

The final group of intelligences is the Functional Group. There are only four ways in which humankind can function: physically, mentally, emotionally, or intuitively. This group describes man in terms of body, mind, feelings, and spirit—how we think, feel, act, and use intuition to express ourselves and obtain our goals. It was Carl Jung who synthesized the views of other psychologists (such as Freud and Adler) while simultaneously augmenting them. Jung concluded that everyone was born with four distinct ways of interacting with the world. These four kinds of interaction correspond precisely with the intelligences in the Functional Group.

Each function also corresponds to a personality type. Studies over the years have shown consistently that identifying our personality type is analogous to recognizing how we are motivated to achieve various results. This premise led Katherine and Isabelle Briggs to develop, in conjunction with career counselors and college placement officers, the Myer-Briggs Type Indicator. This psychological inventory is administered to millions of people annually to help assess how a person functions, which in turn can help determine possible career paths.

The Physical Functional Intelligence

The first Functional intelligence may seem too obvious to be regarded as an intelligence at all. It is the Physical. This intelligence literally refers to matter, the visible manifestation of potential energy. The significance and nature of physical intelligence was never clearer than when I first started with NATI back in 1981. At that time, I was developing a trapshooting enhancement program for myself and some of my business colleagues. We were all heavily into shooting on a competitive basis at the time. Three of these colleagues—Sal Pepe, Andy LaSala, and Steve Giamondo—while top guns on a state and regional level, were seeking to enhance their skills. After only a few months, the results were apparent as these men adapted to NATI processes. Soon, all four of us were finishing in the top five spots in every competition. Basically, we focused on the Human Character Formula. Focusing on *what* we were doing rather than on *how* we were doing was crucial.

It became apparent to me that the Human Character Formula was highly effective in developing physical skills, not only in the area of trapshooting, but also in other physical endeavors such as golf. The amount of concentration focused on the task at hand always has a lot to do with the degree of success realized. This is the Awareness intelligence. In addition, the ability to generate self-confidence, which is the Belief intelligence's part of the formula, also comes into play to a large degree. It never ceases to amaze me, as it did my colleagues all those years ago, how we tend to defeat ourselves by our own limiting perceptions. You can place yourself on the practice putting green and make one three-footer after another without missing a single one. However, approach the same putt during a competitive round and it's like walking across a wooden plank a hundred feet off the ground. It takes on a much greater significance. Success in sports (physical expression) becomes largely a function of how well one focuses on the task at hand, as well as the belief in one's ability to achieve that end. It corresponds with the Human Character Formula.

One of the most amazing things about Physical intelligence and the undertaking of various physical tasks is that the moment one thinks he or she has mastered a skill, everything seems to suddenly fall apart. The 72 on the golf course one day can easily turn to an 89 the next day without any material changes in the playing conditions. For example, professional trapshooters and golfers can perform brilliantly one day and then look like they just walked from an accident on the next. This is true of all physical undertakings.

Let's look at this from the aspect of paradigms discussed earlier. We can embrace new information to develop a larger paradigm, but it is all too common to fall into our old way of looking at things when under pressure. We are very reluctant to fully and deeply assimilate new ways of thinking because it is easier to stay in our comfort zones. The result? Our functioning is governed by older, less productive ways of thinking, and the previous paradigm remains intact. Put another way, the new frame of reference regresses to the old one.

When trapshooting, for example, the brain organizes billions of neurons in its neural network. The greater the ability to concentrate and "image," the better and stronger the organizational capacity of the neurons and their networks will be. There are, however, limits to this ability to organize by any means of standard measurement. Accordingly, I have discovered that the best way to retain focusing power is by adhering to abstractions. Visualizing colors and listening to music are examples. By "imaging," we bypass the short circuits of more restrictive organizational patterns. We are able to embrace the new—and larger—paradigm. Athletes, surgeons, artists, pilots, soldiers, and nurses are all examples of individuals displaying physical intelligence.

Note that this principle relates to one's physical references.

The Emotional Functional Intelligence

To say that some people function emotionally might be considered an understatement of epic proportions, but Emotion is a form of intelligence that goes well beyond any traditional connotation of the word. Emotion may be equated with feeling or desire. When emotion is positive and healthy, it motivates us to achieve an objective and to develop; when it is negative and unhealthy, it often results in failure and restricts development. Unbalanced emotion, for example, may sometimes have a negative impact on any other intelligence we have listed thus far. Tent revival enthusiasm may lead one to a blind adherence of unexamined beliefs. Raw, unchecked emotion may also produce shallow assessments, incorrect procedures, or skewed priorities. Unhealthy emotion can affect relationships, job performance, and self-esteem; on a larger scale it can result in crime, wars, and prejudice. In short, if not used objectively, emotion can cloud our judgment and cause erratic, or even destructive, behavior.

On the other hand, this intelligence carries very positive connotations when balanced. We should not lose sight that motivation and desire have resulted in astonishing achievements throughout the history of mankind. The movement of man from small bands of hunter-gatherers to large populations that use law, reason, and technology represents an evolutionary drive toward potential that is nothing short of astonishing, even when allowing for man's inherent imperfections. During man's long ascent to present-day achievements, man has also had the desire—the emotional intelligence—necessary to produce great works of art, develop medicines, care for those less fortunate, and form systems of government that, with some unfortunate exceptions, have sought justice, fairness, and equality for communities, not just certain individuals.

To say that emotion always clouds reason, therefore, can be shortsighted. On the contrary, emotion is simply one possible avenue that reason can use to express itself. Actors, writers, motivators, caregivers, and parents are some people who utilize emotional intelligence.

Note that this principle relates to emotional references.

The Mental Functional Intelligence

The next functional intelligence is the Mental. This refers to knowledge and IQ, as well as the ability to use reasoning and critical thinking skills. To use a colloquialism, it is the use of brainpower.

In education, the gaining of knowledge is generally a more desirable goal in the long run than the storage of facts (memorization). The mental functionary gets things done by thinking things out versus physically doing them. The cave man threw rocks, while the Romans catapulted them.

Still, if mental skills were to be considered the ultimate intelligence, MENSA members would rule the world, but they don't. As you are learning in this book, human dynamic goes far beyond IQ as we know it. More important than IQ is UQ and PQ—the Understanding Quotient and the Potential Quotient. We all know that people with high IQs are capable of language, math skills, and riddle solving, among many other abilities. But it's the one who *understands* who prevails time after time. It's the one who works on developing his potential who ultimately prevails.

Many heroes from film and literature have been grounded in mental intelligence. Sherlock Holmes, the creation of English writer Sir Arthur Conan Doyle, was able to solve crimes by using a highly refined mental capacity, known in the genre of mystery as ratiocination (or deductive reasoning). To be sure, Holmes' intelligences were razor-sharp, enabling him to amass more information than the average person, but the process by which he solved his mysteries was primarily cerebral. His mental powers and sense of logic might well be said to rival the most sophisticated analytical computers in existence today. (The human brain, of course, *is* a computer, and it is the contention of NATI that everyone can enhance the Mental functioning of his or her "biological PC.")

I contend that people can greatly expand their understanding by using the Human Character Formula, refining their Points (and Frames) of Reference, and adopting the "wholeness approach" to comprehension.

Note that this principle relates to one's mental references.

The Intuitive/Spiritual Functional Intelligence

The fourth and last intelligence in the Functional Group is the Intuitive, or Spiritual. Let me say at the outset that the term "spiritual" as used in NATI does not refer to any kind of religious practice. Associations between overtly religious people and spiritual energy are sometimes made because the ancient practitioners of certain religions seem to have a strongly developed psychic sense. Numerous surveys, however, reveal that people from all areas of life report intuitive phenomena.

To function intuitively is to tap into a higher force or energy that elevates a person's consciousness to a different level of functioning. Such functioning has traditionally been regarded as the sole province of mystics and seers, but this limited view is being challenged more and more by ordinary people exploring the development of their spiritual, intuitive sides. One such person is biologist Rupert Sheldrake. In his book *Seven Experiments Which Could Change the World*, he explains that spiritual growth can be attained by learning to recognize synchronicities.[4]

The term synchronicity, which was first coined by Carl Jung, refers to a meaningful coincidence, one that could not be expected to happen under circumstances governed by mathematical probability. Sheldrake believes that ordinary people can become more aware of synchronicities in their lives by paying closer attention to small events, such as a chance encounter with a friend, an unexpected phone call, the song playing on the car radio—literally anything that comes into their field of awareness. Sheldrake contends that when people pay close enough attention, they begin to notice patterns emerging that represent the intelligence of the universe attempting to communicate with them.

Despite a greater interest in metaphysical literature in the past several years, there are still skeptics who claim that such occurrences of synchronicity are no more than people seeing what they want to see. The idea is really not radical at all, however, if we accept the NATI premise that intelligence is uniformly present in all of creation. If reality itself is intelligent, and we, who are intelligent beings, are a part of reality, then it is quite normal to be in connection with the Whole of which we are Parts. (This is the thesis of Wordsworth's entire canon of poetry, as well as Buckminster Fuller's beliefs.) That we have access to universal intelligence and can have free and easy discourse with a larger field of consciousness is therefore not at all unusual, but rather a normal condition.

NATI never prescribes a particular goal for any of the thirteen intelligences (other than the pursuit of development through virtue, which shall be discussed later), but Intuitive intelligence has many applications. Professional stock market traders balance their economic forecasting skills with old-fashioned intuition. Likewise, entrepreneurs often "go with their gut" when it comes to decision-making. Professional gamblers are usually armed with a great deal of information pertaining to gaming rules and odds, but they also use intuition to know when taking risks might be beneficial.

The connectivity of things we discussed earlier, together with the true nature of Spirit, may both be illuminated by an event that occurred in 1992 in Stockholm, Sweden. At that time, an International Conference of Astronomy declared that over ninety-two percent of the Universe is made up of a substance that science cannot identify. This force, which we labeled as potential, contains within it the notion of Spirit and Intuition. It is more powerful than IQ or anything similar!

Note that this principle relates to spiritual references.

Communication/Reality Reference (CR)
The A+B=C Formula Graphic
POR is the center, then FOR follows.

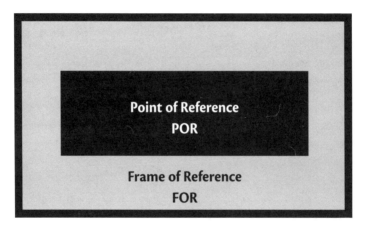

The result (the frame) is CR—Communication Reference.

Chapter Four

Gaining Understanding Through the Concept of Opposition

"The Devil Made Me Do It!"

We have mentioned the concept of opposition, or Polarity, in previous chapters, but what exactly is the role of Polarity in the development of understanding and potential? What role does it play in the mind science that is NATI?

Just as nature presents us with spectacular sunsets and colorful rainbows, it also demonstrates a seemingly callous indifference to life as demonstrated by events such as the tsunami in the Bay of Bengal or Hurricane Katrina in the southeastern United States. It is obvious that evil—or a negative force, if you will—is built into creation.

Theologians and philosophers have grappled with the presence of this negative element for centuries, but NATI simply proposes that it is inherent in the universe and can actually help us develop our potential when properly understood.

The traditional way of dealing with evil is to approach it within the context of religious authority and dogma, which is virtually the only context available to most people because of their upbringing and education. Formerly, there was no gray area; people were either sinful or holy. This concept spilled into other areas of culture as well. At the movies, cowboys either wore white hats or black hats; and balance was only achieved when the black hats were eliminated. But what is the logical result of dealing with the idea of negativity in this manner? Each reader may come to a separate conclusion, but let's consider the matter using the Awareness intelligence. According to Howard Bloom in *The Lucifer Principle*, when awareness shuns certain realities, they paradoxically become part of a process that makes the notion of an enemy very real.[1] The idea of evil clicks!

The idea of a malevolent entity—the devil and the evil he causes—has even

worked its way into the media. Horror novels and motion pictures regularly portray evil as a force that intrudes on our free wills, causing chaos and destruction despite a protagonist's best attempts to thwart it. Evil is always "out there" and to be overcome, not requiring any rational response on our part to what is, in actuality, a polarity within nature. Comedian Flip Wilson made a mockery of our accountability when he coined the famous line "The devil made me do it!" on his variety show thirty years ago.

Unfortunately, people from all walks of life and at all levels of society find it very convenient to place blame elsewhere for their actions, a mindset that has prevailed from the very beginning of civilization. Shamans, soothsayers, and priests of various religions have always been appointed to deal with negative energy and, in most cases, destroy it. Make no mistake about it, however: opposition/polarity exists both in the world and in the make-up of every human being. Acknowledging it, however, is far more productive than denial or rejection.

Two key aspects in achieving understanding and realizing one's potential, therefore, are accountability and responsibility. In order to be physically, mentally, emotionally, and spiritually clear, we must be totally honest with ourselves. As Polonius said to Hamlet, "To thine own self be true," echoing Socrates' earlier saying carved above Apollo's temple at Delphi: "Know thyself." Blaming others for failure doesn't work for intelligent, successful people.

How NATI uses opposition is an important key to understanding and human dynamics, for it is the basis of human interaction and often generates development through feedback, or the Mirroring intelligence. (We shall relate opposition to Mirroring later in the chapter). Indeed, if everyone in the world believed and behaved the same, there would be no way to grow or learn, for we would have no other behaviors to assess or model (again demonstrating with development the intelligences of Modeling and Assessment). The universe would be static and unchanging, and this is certainly not the case. Nature shows us that the universe is always evolving since absolutes equal infinite possibilities.

The mere existence of opposition implies problems, and as we said previously, problems represent an absolute factor in the human condition and lie at the very heart of our journey. Our destination, however, cannot be limited; rather it is infinite, for there is no growth or personal development inherent in the idea of limitation. An easy way to understand this is to regard development as our path, with various problems, weaknesses, and challenges as gateways we must pass through.

Opposition is most commonly thought of as positive versus negative, good versus evil, off versus on. While there may be some kernel of truth in these simple pairings representing opposition, the result is that people regard such opposites in a simplistic manner. They believe the pathway to enlightenment entails eliminating everything that is negative or evil. This either/or mentality, however, is definitely not the way to enlightenment. Simplistic notions of good and bad produce what NATI calls the Great Restrictors: Fear, Ego, Ignorance, and Self-Deception, as well as personality orientations outlined in chapter five. Many of us are conditioned from an early age to think only on the emotional level, driving us to polarities, or adherence to a single way of viewing issues. We learn to fear and loathe "opponents" because of ignorance about the principle of opposition. Our resulting actions are based on the self deceptions we buy into. The visceral, unthinking way we approach marriage, politics, religion, or moral issues are perfect examples of how we don't try to find common ground between "polar issues," or at the very least, understand and work with them. Bumper stickers and flag waving accomplish nothing in the long run.

The way that Natural Thinking & Intelligence approaches opposition is by using the Principle of Complementarity. The principle can best be understood by the paradox of wave-particle duality as set forth by physicist Niels Bohr. Simply put, when an electron has position (is a particle), it has no energy (is not a wave). When it has energy (is a wave), it has no position (is no longer a particle). Contradictory? Not really. Bohr said that electrons transcend both descriptions. The two seemingly opposite states simply complement each other. Bohr ultimately won the Nobel Prize in Science for his Principle of Complementarity, but this same idea of transcendence can be found in both Eastern philosophy (yin and yang) and mind sciences. *The bottom line is that opposition and polarities are a part of nature!* With nature as our model, this principle can therefore be applied to *any* system of opposites to achieve potential, to deal with challenges and evil as we move along the path of development. I discovered the Tao in my personal quest for truth many years ago, and perhaps one of its sayings best explains this concept:

> Long and short complete one another.
> High and low rely on each other.
> Pitch and tone make harmony together.
> Beginning and ending follow one another.
>
> - Tao 2

What could be more simple? To a second grader, the meaning behind "short" is incomprehensible without the meaning of "long," and vice-versa. We can generalize, therefore, and state that the very stability of nature relies on a union of opposites. The resulting energy of this union generates tension, but we must resist the usual connotation of "tension" (anxiety). Tension is a dynamic force enabling creative variation to take place, resulting in change and development, and tension is nothing more than a fight for control between opposites.

Hurricanes are perfect examples of Complementarity, ones that can be found in nature itself. While no one can deny the enormous destruction caused by a storm such as Katrina, hurricanes, like tornadoes and earthquakes, are simply manifestations of the ecosystem and are necessary if that system is to maintain balance and equilibrium. Hurricanes are created by solar energy, elevated water temperature, and barometric pressure. Quite simply, nature must find a path to channel immense energies that build up in the oceans. We cannot, therefore, look upon such natural forces as evil per se. What we can do is to learn better forecasting methods and construct better levee systems and barrier islands to protect ourselves so as to live in harmony with nature. We cannot simply turn off a hurricane, but Complementarity offers us a way to deal with Safety vs. Danger in a constructive way in the matter of severe storms.

Possible applications of Complementarity can be seen everywhere. Another example is seen in the practices of the Swiss engineering company, ABB ASEA Brown Boveri, which employs more than 200,000 people in 140 countries. The company's chief executive, Goran Lindahl, hired a psychologist to survey cultural attitudes among its managers from several different countries. Lindahl contends that his company's profits and high performance are due to the company's multicultural make-up, which researchers say is more likely to produce innovative ideas than a group from a single culture. The tension of Us vs. Them is therefore transcended and transformed into growth and achievement.

Even in the realm of relationships, personal or otherwise, the way to overcome negativity is to transcend it. Loving or forgiving your neighbor does not attain this.

Rather, transcendence is achieved by wisdom, strength (character), and synthesis of all factors concerning a given issue. This last factor of synthesis (one of our organizational intelligences) is better known as turning a negative into a positive (or turning the other cheek). When development is regarded as a transcendent factor, negative experiences are treated as learning events. Love and forgiveness come into play by virtue of not considering anyone unworthy or undeserving just because something negative or unpleasant is occurring. If two people have a heated disagreement, both parties can bypass Right vs. Wrong by looking at the argument as an opportunity to find a third way, or alternative, to solving the dispute. Both people deserve fair treatment and neither is undeserving. By utilizing the intelligence of Mirroring, we can see our own inadequacies and then address our development. Tension can result in a stronger friendship or a new way of approaching some endeavor.

In summary, a perceived weakness can be equated with a lack that needs strengthening. A problem isn't a hurdle; it's an opportunity.

Opposition and Psychology

Psychologists such as Abraham Maslow and Erich Fromm have connected Polarity with issues such as "evil behavior," which we noted is a concept so prevalent in the mindset of most people. As already seen, the question of what to do with the existence of evil is not really addressed in today's society. Maslow and Fromm (and many others), however, have advanced theories that there is a way to eliminate conflicts, struggles, stupidity, jealousy, and many other undesirable personality traits.

To begin, let's focus on Maslow. The problem of pitting good against evil results in what he called unrealistic perfectionism. Maslow believed that the history of the struggle to find a utopia reveals many unrealistic, unattainable, and non-human fantasies, such as "let us love one another; let us share equally; treat all people equally in all ways; nobody should have power over anyone else; the application of force is always evil; and there are no bad people, only unloved people."[2] These aphorisms of goodness, all of which shun the existence of opposites, are unreasonable expectations that, when acted on, invariably lead to failure, apathy, disillusionment, discouragement, hostility, and so forth. How can they not, for they are rooted in an incorrect view of the design of nature.

Other scientific and psychological theories further emphasize the necessity of Polarity in development. Besides Complementarity, the stability found within nature is also based on the union of opposites. Matter itself is built on the relationship between positively and negatively charged particles (protons and electrons), which we all learned about in high school. The growth of vegetation results from both sunshine and rain, or Dry vs. Wet. The violent explosion of stars in deep space creates chemical elements that will form new stars and planets (Destruction vs. Creation). We think of summer and winter—hot and cold—as opposites, and yet the seasons are crucial in regulating the environment and making it habitable. Everywhere in nature tension creates variation and change that results, not in chaos, but in stability and equilibrium.

Even within the ecosystem, man himself can adapt to almost any environment, but he is rarely satisfied. This dissatisfaction resulted in evolutionary development and momentum (the use of a "negative" to develop the species). Over tens of thousands of years, man moved into both polar and tropical climates and adapted, creating the cultural and ethnic diversity of life we see all around us. In fact, brain research indicates that this structural framework of tension is characteristic of such adaptive responses. Humans *need* both predictability and variation, security and independence. We are geared toward finding shelter and protection, but we are also explorers by nature, as evidenced by human expansion across continents and a drive to land on the moon. Without polarities—or the tension inherent in opposition—there is no cultural, scientific, spiritual, emotional, or mental progress. We can say that human nature is patterned (or modeled) for the tension that makes change possible, and Polarity promotes that patterning.

Precisely because we are a part of nature, life is always placing us at the crossroads of Polarity. Whenever we employ a plan, we are usually choosing against its opposite. In a very real sense, life pushes us to make such choices, and choices are often difficult and painful, but they are also opportunities to grow and evolve as we contend with the consequences of any given situation.

Polarity is also part of our inner being. As Myer and Briggs definitively established, we experience each action from two perspectives: one from the person we are inside (our introverted view), and one from the person that interacts with others (our extroverted perspective). The two perspectives are not exclusive, but rather symbiotic in nature. Jung explained this concept by saying that our two personalities are similar to two chemicals: when they meet, both are transformed. The original components are still present, but they become synthesized into a whole. This, and all transformations occurring as a result of tension between opposites, is transcendent in nature.

Speaking in terms of the dynamics of personal relationships, Fromm believed we often have strong preferences for one attitude because we are paradoxically drawn to its opposite. The affinity to this attitudinal opposite is natural and leads to relationships in which both parties are able to change and grow. This occurs when partners are willing to listen to advice or observe examples of a behavior that reflects the opposite attitude (or their secondary functioning, if you will).

These psychological principles are in perfect harmony with the intelligence of Mirroring. When people overdevelop a dominant attitude, they become attracted to those who will actually reflect the *opposing* attitude. Another person can put us in contact with our unconscious side while sparing us the ordeal of reconciling our own contradictions. In the initial stages of such a relationship, people generally experience a sense of well- being. Over time, the connection inevitably generates tension. Fromm believes that both parties then depend on each other for a sense of wholeness, with individual growth threatening the arrangement and possibly destroying the original sense of well-being.

This does not imply that symbiotic relationships are unhealthy, but it is NATI's contention that when we have the strength of character to "look into the mirror," we can reconcile our contradictions and grow, achieve, and develop without threatening the relationship.

Opposition can lead to an infinite number of exciting and interesting patterns of behavior and personality. The social sciences are replete with theories that are synonymous with the psychological dynamics described in this section. Natural Thinking & Intelligence believes that its principles are the basis for the collateral identifications of these dynamics.

Using Abstraction

Theory is all well and good, but as we noted, we have been raised to look at the world with an either/or mentality, and trying to overcome such an ingrained view of the world is extremely difficult. We know that electrons—matter in the most conventional sense of the word—behave as both solid objects and energy. But how can that be?

Perhaps we can take our cue on accepting opposites from someone who typifies the Mental intelligence, none other than Albert Einstein, and we can do so without getting bogged down in lengthy discussions of physics. We will simply state that Einstein saw that matter and energy were two aspects of the same thing—his Theory of Relativity (or $E=MC^2$. Even Niels Bohr, the famous physicist who framed wave-particle duality and the Principle of Complementarity, said that he found it exceedingly difficult to imagine this kind of duality.

So consider this: the Theory of Relativity—that the universe is both matter and energy—was formulated by Einstein after he had been lulled into a daydream by staring at dust motes swirling in a sunbeam during his lunch break. (Einstein often found himself caught up in such daydreams while working as a lowly clerk in a German patent office early in his career.) This actually correlates with some personal experiences of being able to translate the abstract to the concrete after imagining colors or listening to music. There is no precise formula, of course, for using one's imagination. Such exercises are much like mentally "letting go," so that our minds can function on a more intuitive level for a while. It's like giving oneself a little mental breathing room. It is no accident, perhaps, that people who learn how to let go of the noise and clutter of everyday life through transcendental meditation are able to find greater insight and understanding in their lives. Remember that putting opposites together is inherently a *transcendent* activity. I'm not suggesting that we look for a guru, and indeed, NATI says that the real guru already exists—inside ourselves!

So where does that leave us?

What I am suggesting is that we can find abstraction, which is responsible for unification and wholeness, by tapping into our creative, intuitive sides. Since we all possess the Intuitive intelligence, this is not as difficult as you might think. If you are faced with a problem, you are faced with a choice between two opposites—the crossroads I mentioned a moment ago. Try painting, dancing, playing music, walking, exercising, or any activity that you find kicks the imagination into gear. I have always found that once people do this, they are better able to frame an issue. Imagination is a transcendent tool, for it represents what might be, not what is—and that is Polarity! Once you get into the habit of accepting such imaginary exercises, you might find that your mind and brain can synthesize an answer to your problem. You may not figure out the Theory of Relativity, but you might enjoy better health or relationships that are problematical precisely because they present you with the tension of opposites.

We started this book by talking of scientists who regarded the mind and brain as two different things—the universe is either "in here" or "out there." This is a classic case of duality! Understanding Polarity enables us to expand our minds in ever-creative ways in the service of growth and the achievement of potential.

So when you consider the ideas in this chapter, use both your Mental and Intuitive intelligences. They may seem to represent opposites, but used together, they can achieve startling results!

Start out by asking yourself whether you are really energy or matter.

The answer is simply "yes."

Reversing Polarity

Here is an example of Polarity/Opposition and how it was overcome. Around 1980, I decided to go after major corporations as clients for my real estate tax consulting business. I appealed their property taxes and had them reduced. The problems I encountered were: I was not an attorney; I didn't have a track record of success: and I was an unknown entity. Here's what I did to offset those oppositions:

1. I hired a public relations firm to promote my small, local successes and used the PR to establish myself as a known entity.

2. I took a specific property owned by Ford and did a totally comprehensive analysis to show the merits of my analysis of a strong tax reduction.

3. I asked a real estate tax attorney to write an opinion as to the merits of my analysis. I also asked him if I could retain him to handle the legal aspects only of these matters which he acknowledged.

I then approached Ford and Chrysler with my proposal. As part of the proposal I demonstrated that attorneys are limited in the areas they serve while I was not. Also, attorneys take four years or so to finalize these matters. I showed them that I could do it in two years. The result was that I represented Ford for twenty-three years in seventeen states and Chrysler for seven years on a national level.

The polarities were completely and successfully reversed!

Chapter Five

Building the NATI Thinking System

As stated earlier, there are certain segments of our NATI structure that deal only with the thinking system. This chapter deals with those segments. It is also important to note that segments of this type are more important for understanding the NATI concept than its implementation.

Core Human Dynamics

Thus far we have covered principles of nature, otherwise known as innate human intelligences. This is the technological aspect of decoding potential and understanding. We will now address the thinking, or dynamic aspects, of the equation, of which there are several. They are Core Human Dynamics, the Great Restrictors, and Personality Orientations. These are the factors that give us our character and uniqueness.

In the organizational network previously described, we noted that company employees possessed characteristics such as drive, motivation, acceptance, uniqueness, and good judgment. NATI believes that characteristics such as these are fundamental to the human condition. There are quite a few, but NATI identifies the following as the most important dynamics responsible for human behavior. We call them the Core Human Dynamics. They are:

- Control (influence or regulation)
- Power (potency, strength, or energy)
- Inclusion/Acceptance (being a part of something or containing other parts)
- Exclusion/Uniqueness (restrictive or reluctant to accept)
- Attention/Recognition-seeking (calling attention to oneself or being notable)
- Self-interest (interest in personal advantage, motives, or comfort)
- Judgment/Values/Appraising (opinions, valuations, conclusions)
- Drive/Motivation (to act, instigate, or desire an action)

Core Human Dynamics (CHDs) should not be interpreted as either positive or negative, but as neutral. In any given situation they may produce healthy or unhealthy behavior. Timothy McVeigh, for example, certainly felt that people who shared his beliefs were being excluded from society. He therefore made certain appraisals of situations (events at Ruby Ridge and Waco) and then sought control through power. As we can see in the corporate model already cited, however, these very same core dynamics can be used in very positive, productive ways. My own experience is that people have an overabundance of one or more of these in their personality.

Core Human Dynamics combine with basic NATI principles to form recognizable mindsets. For example, if one focuses on a

Concept or Belief concerning control, he or she will naturally and automatically consider the following:

- rules about control
- priorities regarding control
- integrating control
- the measure of control
- the process of control
- details of control

The same process would naturally apply to other CHDs:

- rules about power, exclusivity, judgment, etc.
- priorities regarding power, exclusivity, judgment, etc.
- integrating power, exclusivity, judgment, etc.
- the measure of power, exclusivity, judgment, etc.
- the process of power, exclusivity, judgment, etc.
- details of power, exclusivity, judgment, etc.

Moreover, by combining the CHDs with the Functional categories, a more in-depth understanding of people's mindsets and actions is revealed by considering their:

- rules concerning physical, mental, emotional, or intuitive control
- physical, mental, emotional, or intuitive priorities regarding control
- physical, mental, emotional, or intuitive integration of control
- physical, mental, emotional, or intuitive measure of control
- physical, mental, emotional, or intuitive processes of control
- physical, mental, emotional, or intuitive details of control.

By following the above models of the various NATI principles, we can readily ascertain an accurate life philosophy. Core Human Dynamics are very important because potential is realized through your mindset. Your mindset is what you are—the very seat of your soul—and what you are at present naturally determines what you can become as well as your overall purpose in life.

In hindsight, we certainly know what kind of rules, priorities, beliefs, and patterns of control and power dominated the mindset of Timothy McVeigh. His radical patriot mindset, hindered by the Great Restrictors, ultimately led to certain mental, emotional, and physical functions relative to CHDs. Productive corporate strategies, using the same CHDs, led to entirely different functioning. The former mindset represents the quintessential closed system. The latter represents the open, continually evolving system.

The Great Restrictors

Our lives are so fraught with tension that the sympathetic nervous system is frequently stuck in overdrive for much of the day. Evolution has predisposed us to the Great Restrictors of Fear, Ego, Ignorance, and Self-deception. In essence, we are genetically hardwired for these, which is another way of saying that evolution has worked both for and against us over the millennia.

Unfortunately, the Great Restrictors are the antithesis of development, so that man is frequently a product of rote, mechanical thinking. In areas of culture, government, education, and religion, to name just a few, we use an emotional approach to problem solving, one that is based more on the chemicals produced by the reptilian parts of our brains than on logic or reasoning. Paradoxically, this instinctive approach doesn't solve many problems at all in the long run, although we do succeed all too well in degrading our social systems while eroding our personal health in the process. Mind-body medicine has shown clearly that stress caused by the Great Restrictors is responsible for much of our ill-health, especially heart disease, cancer, and compromised immune system functioning. A vicious circle is then created (a closed system, if you will, with near-lethal feedback) whereby compromised health impairs our ability to function at higher levels or to use Directional Judgment to achieve virtue.

The consequences can be seen everywhere around us. For instance, while virtually every American believes that we live in a democratic environment, a limited number of citizens are willing to work in the support of a free and open society. In the 2000 U.S. national elections, only forty-eight percent of eligible voters participated, and turnout for local elections is usually even worse. We seem to have lost sight of Jefferson's timeless admonition that "The price of liberty is eternal vigilance." To believe that our society will continue to preserve freedom given such apathy in the general population is the height of arrogance and self-deception.

Consider the following facts. We are rapidly approaching the point at which fifty percent of the work force will be employed by one of the hundreds of governmental agencies that have sprung up within our nation's bureaucracy. Theoretically, this group could decide to vote *en masse* to effect a change in political or economic policy, with the other half of the electorate having no choice but to go along with the block vote of the cohesive fifty percent. In many ways, this is already happening. The salaries of some local, part-time politicians have been set by a handful of voters, even though the salaries are totally out of proportion to the work performed by those in office. By the same token, activists and lobbyists now seem to have far more influence in governmental policy than scholars, intellectuals, or reformers. A minority is influencing the decision-making processes, which belong in the hands of the majority. Perhaps the most frightening aspect of this concentration of power in the hands of a few is that people have grown indifferent towards political processes. They have become discouraged, believing that they can no longer make a difference in the face of Big Government. (Behind this fatalism is a restrictor, namely ignorance of the fact that America is a government by, of, and for the people.) This attitude of skepticism is based on emotion, not on an understanding of how the Constitution provides for a self-governing population.

Another area in which the unhealthy emotions of indifference and apathy operate is in education. Elemental subjects such as World History have either been eliminated or altered by concessions to political correctness; discipline is almost non-existent in urban, middle-grade classrooms; and teachers are cowered into submission by student threats or school populations that must pass through metal detectors each morning. In college, the only requirement for a degree is paperwork, the price of admission, or athletic prowess. The result is that both values and academic performance continue to decline in our nation's schools.

As far as religion is concerned, the United States has always been described as a God-fearing nation, so it is no surprise that ninety percent of Americans profess a belief in God. Sixty percent of these belong to a formalized religion, and yet only half of this number practices their faith in any significant manner. Of the remaining thirty percent, there is an alarming number who do not even understand the basic tenets of their faith.

The picture is no brighter when the entire continuum of culture is examined. Corruption in business is rampant, as evidenced by the recent Enron scandal and others. Radical subcultures are evolving at an alarming rate, ones involving cults, mind control, suicide pacts, drug use, gangs, sexual abuse, and hedonism. Reality shows have reduced sex and love to strategies enabling contestants to win millions of dollars. The shows' premises are based on deception, illusion, and rejection, and yet they pass for entertainment, with most viewers never questioning the values that are the foundation for the recent "reality craze." As Burt Laurel of Laurel and Hardy says; "This is another fine mess you've gotten us into, Ollie."

Personality Orientations

We have talked a great deal about the transcendence made possible by Complementarity in the last few pages, but I want to emphasize that NATI does not require anyone to adopt specific religious or cultural beliefs that are sometimes connoted by this term (transcendence). While using hard science and respected philosophical traditions to explain various concepts, we are ultimately concerned with helping people do things more smoothly and with minimum stress, hardship, and worry. The transcendence we strive for is concerned with accountability, responsibility, understanding, and development, not with doctrinal beliefs.

To this end, look at the list below, which represents some of the many Personality Orientations that represent polarities in our everyday activities and relationships:

potential   =   possibility
actualized =   done or accomplished

introverted =   inward-oriented
extrovert   =   outgoing

open   =   willing to accept, expansion
closed   =   unwilling to accept, restricted

objective   =   impersonal position

subjective	=	personal position
variant	=	focus on dynamics, change, and flexibility
invariant	=	focus on fixed and static patterns
self	=	self-centered
selfless	=	other-centered
rote	=	doing things automatically
thinking	=	examining objects and ideas carefully
abstract	=	relative, archetypal
concrete	=	factual, tangible
positive	=	plus
negative	=	minus
defensive	=	guarded
aggressive	=	forward
weak	=	lacking
strong	=	capable
probable	=	likely to occur
definite	=	determinate

Weaknesses

Let's look at some fascinating repercussions caused by the Personality Orientations cited above. Keep in mind that Personality Orientations are patterns adopted as a result of one's Beliefs and Focuses. As such, when we take these for granted, we lose the ability to satisfy the deep, driving facilities that inexplicably take over our psyches and generate poor patterns. The following are just some metaphors we utilize for our students in identifying their orientations.

- **Structured**–If things don't work in a structured manner, then I get lost and can't find my way.

- **Unconcerned**–I will just stay outside of things because if I get involved it will look like I am on the same level as others. Additionally I may learn something about myself that I don't like.

- **Noninvolvement** – If I don't deal with the question, I don't have to face its meaning.

- **Superficial Ways**–Everything is fine, so don't go any deeper; it may ruin my style.

- **I Am Different**–If I am different, then I have to conform and face up to what is wrong, or my weakness.

- **I Know**–Even though I cannot answer your question, I know, because if I didn't know, everybody would think I am stupid.

- **The Holder**–I hold things in and sometimes I want to shout or hit people, but I let it pass because I cannot express myself, and I don't want to be hurt by confrontation.

- **The Violent Reaction**–I will attack you because I cannot tolerate the pain of being hurt emotionally, nor do I have the mental ability to deal with you.

- **The Attitude**–I have so many things going for me that I block out everyone and everything, including my inner voice.

- **The False Air**–What I am saying is actually the opposite of what I mean because I am afraid you might see the real me.

- **The Green One**–I don't see anything better, so I go for everything I can. After all, that is the name of the game, and money is the most important thing in life.

- **The Egotist**–What you don't understand is that I am the center of the universe. What *I* don't understand is "so is everybody else."

- **Playing Dumb**–I always miss the point so I don't have to deal with the issue.

- **The Advocate**–I simply cannot agree or be objective because I have too much to lose.

- **The Wise One**–Something just happened that made me realize how dumb and naive I am. Now I want to show everyone else how smart I have become.

- **Frustrated and Restricted**–I can't develop my potential under these current environmental patterns. However, I am too afraid to break out of them.

- **Please Reject Me**–A lack of confidence, or a "lost" mindset. I pattern myself so I will eventually be rejected. I become self-rejecting.

Every one of us is composed of a multifaceted matrix with strengths and weaknesses. By identifying and addressing these weaknesses, you can reach virtually any goal you choose. When you bring focus to bear on weakness, you can, among other things, determine what NATI calls your Current Operating Procedure, or COP. COPs can be unhealthy behaviors that corrupt everyone's naturally clear and efficient matrix at some point by rigid, institutional thinking. By becoming aware of your COP, you can learn how the Great Restrictors hold you back to one degree or other. It makes no difference whether you want to improve your golf game, quit smoking, be a better student, or enjoy healthier personal relationships. Identifying your COP can help improve your life. We will discuss this subject later at length.

Always remember that we should never be afraid of weakness.

You may therefore recognize traits in the above list, aspects of yourself that you would like to change. You may be afraid to try new things and listen to the views of others. You may be introverted and passive, fearful of expressing how you really feel. Perhaps you are weak and defensive in certain situations, always on guard against what you perceive as possible harm. Maybe you're not aggressive enough on the golf course. If you're trying to quit smoking, you may feel that the chances are stacked against you so that your attitude is "probable" rather than "definite." Regardless of what weaknesses you may have, you can use them to grow by first becoming aware of them and then examining your beliefs about them, such as:

- Am I afraid to listen to others because they think they are smarter than I?
- Do I withhold my point of view for fear of being ridiculed?
- Do I think I will be hurt in a relationship because my previous partner treated me with cruelty?
- Am I worried about what others are thinking when I swing a golf club?
- Do I think any attempt to stop smoking is doomed because I tried before and failed?

Your Character, or what you are capable of becoming, is *always* equal to your Awareness and Belief: A + B = C.

Your thirteen natural intelligences are fixed and unchangeable, but how you use them is a different matter. You are a dynamic system capable of infinite change and development. Pursuing your development through potential can help you to transcend your weaknesses. As we saw in our many examples, energy can be redistributed when you apply your intelligences and become aware of Polarities. You are an open system in which anything is possible! As William Blake said, "We are all creatures of energy, and energy is pure delight."[1]

"Potential Exploding"

Chapter Six

Potential

"92% of the universe is composed of a substance we cannot identify."

—astronomical fact

Potential is the most important word in the entire NATI philosophy, for it represents the possibility of reality. It is an unsatisfied, undeveloped universal force that seeks to be realized or expressed. Potential is latent and can be manifested in an infinite number of possibilities. It represents all energy, matter, and life that have existed in the past and present, or will exist in the future. Potential brings forth all forms of life and is one of the few absolute principles of existence. In this respect, it is all-encompassing since it contains all possibilities.

The scientific definition of potential is "energy stored in an object, or the ability of an object to do work." For instance, a rock at the top of a cliff has more potential energy than one on the ground because of the force involved in getting from the cliff to the ground, thus converting the potential energy to kinetic energy. The notion of energy stored in an object represents a contradiction in terms to potential as a force. It assumes that it requires an object to take an action. This, of course, places greater significance on the object than it does on the object's potential.

This presents an interesting issue concerning the original nature (or "first cause") of potential. In an ontological sense, one might say that potential emanates from the force known as the Absolute. This notion, however, leads to the age-old philosophical complaint that the nature of God could not be personal since it would also have to be equally positive and negative. Nature, however, contains all possibilities, including negative ones, and a humanistic god is, according to traditional thinking, not negative.

Potential and Probability

According to probability theory, if a golfer has missed a five-foot putt eight times in a row, the chances are that he will miss it on the ninth try, just as a salesman who has pitched a product eight times is more than likely to miss on his ninth attempt. At some time in this dynamic, however, something is going to happen whereby both the golfer and the salesman are going to succeed. The golfer is going to sink the five-footer and the salesman is going to make that coveted sale. It is at this point where something occurs in the dynamic whereby the probability is reversed. We can say that the odds are now in favor of the desired outcome being realized. According to my experiences with Natural Intelligence, what is really occurring is nothing less than a shift in consciousness. This happens according to two of the principles of Natural Intelligence: Belief and Emotion. In other words, confidence and determination are finally being used by the golfer and the salesman! Without awareness, this shift in consciousness is extremely subtle in the sense that we do not realize that something is occurring within us. Indeed, what we may be sensing is the anticipation of potential realization. We are then connecting with the field of potential.

My research has shown that the anticipation of potential realization is really an intuitive impulse, or a preparation for the eventuality that is developing. This includes a pattern shift, which is attributable to Awareness, Belief, and Emotion, three of the NATI intelligences. On an unconscious or psychic level, the golfer says, "To hell with this. I'm sick and tired of blowing this simple putt. This one is going in." He then becomes convinced that he can do it and generates an added amount of desire towards that end. He then overcomes his fear. Hence, he accomplishes it. The same holds true for our salesman. In effect, the quantum field of potential collapses. In simpler terms, this collapse involves energy components (known in science as quanta) that make up a possibility. Quanta are bits of energy that carry data. The reason the field fails, as with other cases of failure to accomplish a goal, is that people are focused on the Great Restrictors. In the two examples of success, however, a paradigm shift has occurred. The energy of the restrictors is transformed into a positive probability field in which the wave does not fail. Why and how does it occur? It is simply due to the expansion or elimination of our restricted awareness, or focus, resulting in an enhanced belief in one's ability.

Let us also look at this issue in terms of an earlier example. If I ask you to walk on a plank two feet wide and four inches off the ground, I doubt you would have any problem completing the task. If I take that plank and suspend it one hundred feet in the air between two buildings and ask you to once again walk its length, the scenario changes dramatically. Why? It's the exact same objective, isn't it? The answer is that your awareness and perceptions have added the factor of fear. What then happens is that there is a shift from a field of predictability ("I can make the crossing") to a field of low probability (No way!"). The field of potential collapses. The energy of desire and confidence is altered by fear. In effect, our Point of Reference and our Frame of Reference are altered, thereby causing a paradigm shift in the Potential/Reality fields.

This is what occurs with the functional aspects of any undertaking. We have perceptions and awareness, or the lack thereof, within our models and paradigms, and these govern our functional success or failure. In the case of walking the plank at a higher altitude, the probability of failure is increased by the fear of losing one's life.

An example of how NATI deals with this type of scenario concerns my daughter, who was a competitive ice skater during her teen years (a really good one, winning a number of competitions). Like so many athletes, fear and doubt caused a case of nerves before her performance. During one particular competition, she fell doing her short program. While this is not uncommon, this part of the program was one of her greatest strengths on the ice, and she did not score well because of the fall. Afterward, she came to me quite dejected and asked what I thought the problem might be. I told her we would discuss the matter before her next competition, which was the following day. Several hours before the ensuing competition, she brought up the subject again and I told her that there are times when ice skaters (and athletes in general) put themselves in a frame of mind in which the results are so important that simply walking across the ice would be nearly impossible if asked to do so. Her response was "Yes, that's me alright!"

Simplistic or meaningless? Not in this context. My comment was a powerful observation at a critical time. It made my daughter realize that on a physical level she was placing significant emphasis on the loss of a skill (and the competition it was part of). But loss of skill is also loss of effectiveness, effectiveness being a quality of nature we set forth in chapter one. Stated differently, she experienced a lack of confidence. My discussion with her clarified the issue by making her aware that *all* the competitors were undergoing similar states of consciousness to a greater or lesser degree. It was this state of consciousness (lack of confidence) that was going to be the determining factor in the field of probability regarding their achievements. The effective NATI personality in this instance is one who is willing to give up the self—his or her point of reference, or the "I" part of the equation—and become connected to another paradigm, that of self-development. I have repeatedly found that when one focuses on developing potential rather than the self, the achievement level is much greater. One is freer to be who one really is—or can be.

Types of Potential

There are five types of Potential:

- Absolute Potential (AP): The underlying universal force that seeks to be actualized
- Ultimate Potential (UP): What a person can do or have; the amount a person can achieve
- Desired Potential (DP): What a person wants to do, have, or maintain
- Realized Potential (RP): What a person does and has
- Unrealized Potential (UNP): What a person can do and have, but doesn't

Using this simple formula, you can gauge your unrealized potential:

Potential Principle UP – RP = UNP

The key here is how is to determine one's UP in order to see what still needs to be developed, or one's UNP. The basic premise is to follow the Human Character Formula of A + B = C. The ultimate determination is based on the degree of energy or spirit we contain, the manner we utilize the Core Human Dynamics, and adaptations of Personality Orientations in conjunction with the Human Character Formula.

The Field of Potential

Some time ago, Dr. Langham stated, "Before energy comes *potential*." I have since come to learn that potential is composed of creative, organizational, and functional components. This leads us to the field of potential

Decoding potential is a great deal more than what most people imagine. It involves information and discoveries about how we as humans are representations of pure potential. *Decoding Potential* describes potential as unrealized energy that contains various forms and levels of consciousness. In this regard, Book I dealt with the science of potential. In order to realize the optimal meaning of Book I, it is recommended that we look at why things do or don't happen in our lives. This observation, however, should not occur in a judgmental or negative way (nor in an isolated fashion). Instead, we should examine basic, absolute principles with a view toward understanding and testing our internal, innate, natural structures (which some might call the soul). Examining these factors is necessary, for until we understand ourselves and our basic structures within the cosmos, we can never truly realize our potential. We cannot fully analyze ourselves, effectively communicate with our fellow beings, nor relate to our Creator in a clear, meaningful, and objective sense without an understanding of the self. So where do we find the answers required to achieve the above? Where do we find our potential? The answer is "in nature," which comprises the laws, principles, and structure of reality.

This, then, remains our ongoing focus for Book II: What is potential; how do we fit within potential; and how we use (or fail to use) it? Notice I said "how we fit within potential," not how potential fits within us. Most readers might assume the latter would be a more logical approach, but that assumption would be based on how we normally observe things, not the natural order of the universe. First comes potential, and *then* comes humanity. We evolve from consciousness, as the metaphysicians say. Put another way, when the awareness factor of potential is ignited, we evolve. That having been said, this book is not necessarily going to tell you that the way to make more money is by following this or that principle (although that can easily be a happy byproduct for some). Instead, it utilizes the concept of a potential field and pragmatically explains potential and its manifestation, or lack thereof, within humanity.

Applications and the Mechanics of Potential

The field of potential exists always and everywhere. It can never be removed. It is the carrier of all possibilities. It is the void out of which the proton creates other sub-atomic particles. Mystics express the same idea when they call the ultimate absolute reality Sunyata—the void. They affirm that this is a *living* void, which gives birth to all forms in the observable world. Taoists called this infinite void "pregnant." Their belief is that the Tao of heaven is empty and formless. They do not, however, mean this in the typical sense in which we understand "empty." Indeed, this void has infinite creative potential. The energy of the Tao responds in various parts of the field of potential in which it is activated. It comes together in order to manifest its pregnancy.[1] This energy is exemplified not only in physics (as quanta), but in Chinese philosophy. The field is not only implicit in the notion of the Tao as being empty and formless (yet producing all forms), but is also expressed explicitly in the concept of chi, or the life force.[2]

A crucial feature of this field theory is the creation and altering of paradigms. This will be important in human problem-solving applications and understanding for the following reasons: the creation and alteration of paradigms demonstrates that such a process can be conceived only when paradigms are seen as destructible objects. In other words, they can change. Instead, they are seen rather as dynamic *patterns* involving a certain amount of energy, which can then be redistributed with new patterns forming our reality. This is merely another way of stating that reality and potential are complementary—two expressions of seemingly opposite phenomena. By analogy then, how often are new undertakings and ideas embraced with enthusiasm, as surges of energy that represent transformations from potential to a tangible goal? In terms of quantum physics, the creation of a massive particle is only possible when the energy corresponding to its mass is provided.

Potential energy, therefore, is always responsible for the creation of objects in the observable universe. In the same way, potential energy as defined by NATI, is responsible for the redistribution of energy to form new patterns of achievement. Because nature is our model, we see the principles of science reflected in the NATI philosophy of using energy from the "pregnant void" to manifest itself according to our goals. The transformation is achieved through our Focus and Beliefs, which help actualize the Expression of the initial energy (or enthusiasm to create or change something). Again, A + B = C.

One of the truly amazing phenomena here relates to how this use of energy is sometimes strong and at other times weak. The reason for this difference is explained by Bohr's Principle of Complementarity and Heisenberg's Uncertainty Principle. In the field of potential, all interactions are pictured as *virtual* particles. Virtual particles are different from real ones that are created in the activation process because they only exist as probabilities. They are realized by virtue of observation. Observation equals Focus! Again, we are brought back to Awareness, or our point of reference, as the factor determining the strength of the energy required to realize potential. As we saw with my daughter, it can be diluted by a restrictor such as fear, or enhanced by belief, such as confidence.

It all comes down to the following: potential, as we understand it in terms of both physics and mysticism inherently possesses a quantum field that is seen as the basis of all reality, abstract and concrete. Potential, relativity, and the quantum field are therefore synonymous with respect to human consciousness. Since they exist everywhere, they can never be removed and are "translators" for the energy that produces all material phenomena. Further, as stated by the Chinese Sage Chang Tai, when one knows that the great void is full of chi, one realizes that there is no such thing as nothingness.[3] Potential embraces (or more accurately contains) quantum energy. When quanta (or A + B = C) are generally aligned, events manifest. When they align *perfectly*, the highest degree of energy or power manifests. This latter energy is identified as spiritual or superhuman.

Human Adaptation of the Characteristics of Potential

Within each human, there are certain characteristics that empirical research has demonstrated have been adapted from those of potential itself. The absolute qualities of potential are as follows:

- Omnipotent (all-powerful): to do/influence/control everything.
  - o Positive Aspect: possessing the strength to overcome adversity to develop/evolve.
  - o Negative Aspect: controlling others against their will/ abuse of power.
- Omnivorous (all-engulfing): to have/possess/acquire/consume everything.
  - o Positive Aspect: abundance
  - o Negative Aspect: greed/over-consumption/gluttonous/ short-term/hedonism.
- Omniscient (all-knowing): knowing everything.
  - o Positive Aspect: curiosity/knowledge-seeking
  - o Negative Aspect: jumping to conclusions/ closed-minded / dogmatic / prejudiced
- Omnibus (all-inclusive/all-encompassing): being everything
  - o Positive Aspect: unified/comprehensive/connected
  - o Negative Aspect: lack of uniqueness/distinction among the parts/rejection of all-inclusiveness
- Omnipresent (present everywhere; immortal): to be everybody and always be
  - o Positive Aspect: mobility/longevity
  - o Negative Aspect: overextended/fear of death
- Equipotent: (seeks balance from high to low and vice versa): balanced/homeostatic/ comfortable

o Positive Aspect: moderation/balance/comfort

o Negative Aspect: resists change/stagnation/boredom/low frustration and tolerance/maintains the status quo even when it is negative

Let's now look at specific examples of these characteristics.

Omnipotent

The Omnis have profound effects on people but are greatly misunderstood. For example, take "omnipotent" in a spiritual sense. Fundamentalists believe God has the ability to control us and change our minds. If this is the case, why then do we have dictators, devils, bullies, bosses, boards of directors, organizations, governments, and office holders? On a positive note, however, having inner strength and/or confidence is a representation of omnipotence. In another respect, this is where our core human dynamics of power and control are based. The arrogance of power is bred here.

Omniscient

This is an all-knowing factor we have all experienced internally and externally, both from ourselves and from others. This is the realm of the "know it all" in its weak aspect, while wisdom can be interpreted as its strong respect. At the highest level, it is illumination, or pure knowing. Not many people experience this last state because they don't look in the direction of understanding, development, or truth.

Omnibus

This "be everything" aspect of potential is very exciting and quite revealing. Imagine a space where every possible reality exists.

We may wish to call this space infinity because of the infinite possibilities it presents. What this space needs to make sense (or be realized) is a structure of some sort. The nature of the structures we choose determines what the end result (or manifestation) looks like. Many times we get carried away with what we want to have included in our structures because this segment of potential has overwhelmed us concerning a particular objective. The "I want to do everything" types have an overabundance of this characteristic.

Omnipresent

This segment generates "I am immortal" feelings. It is also related to wanting to be everywhere. It is where our fear of death has its basis. It relates more to space than time, although in its purest state it really doesn't deal with time at all. Therefore, when the reality of time is introduced into this space, conflict arises. Take, for example, travel to exotic places. We all want to go to the mountains, the shore, or the desert, but money aside, we can't be in all these places, especially not at the same time or timeframe (and in most cases, not even in a lifetime).

Omnivorous

Did you ever meet a business person who is consumed with making money? Ivan Boesky couldn't make enough even though he accumulated billions. His greed so consumed him that he transcended the structure of reality we call business and civil laws. That's because he was overwhelmed with omnivorous potential and his structure was not sound. He needed a better, stronger model to follow in order to contain his overabundance of this type of potential.

Equipotent

This represents the middle ground, or balance. Sometimes erroneously cited as being pacifist in nature, it actually represents as much a plus as a minus, strength as much as weakness. This aspect has more to do with non-motivated states, such as willingness to take second place. Psychologically, this may be interpreted as seeing both sides of the issue, or construed as ambivalence in other contexts. Satisfaction is a form of this factor, a very important one as we will see later.

I'll repeat the past comments on the essence of segments, such as those presented in this chapter: it is more significant that they bring the reader to as full an explanation of what we are talking about as possible rather than to attempt to utilize them.

Summary – Doctrine of Potential

Potential is something that does not yet exist but can be imagined. It exists throughout nature. There are two basic types of potential: Realized and Unrealized.

Keys to Potential:

- Realized potential emanates from a point of reference.
- Unrealized potential has no point of reference.

Both forms of potential have three basic components which contain all possibilities:

Creativity – Organization–Function

Furthermore, both forms of potential contain Polarities:

good/bad
positive/negative
on/off
in/out

The basic components have thirteen parts to them:

Creative	Aware/Focus	
	Beliefs/Concepts	
	Expression/Communication	

Organizational	Laws	Mirror
	Order	Details
	Priority	Whole

Functional	Physical/Material	
	Mental	
	Emotional	
	Intuitive/Spirit	

Chapter Seven

A New Philosophy of Understanding

As we have stated earlier, decoding one's potential has two basic parts. The first deals with the "intelligence" portion of NATI. It is science, or the demonstration of nature's principles that are described as intelligences, of which there are thirteen arranged into three groups. The second deals with the "thinking" portion of NATI. It is the philosophy of comprehension, or the explanation of how to utilize NATI principles and how to organize them into a systems approach of comprehension.

We have already discussed many NATI doctrines. In this chapter, we will now discuss the following:

- A general, all-encompassing system of comprehension
- Development as an ongoing mission
- Organizing data
- Development as mankind's underlying force
- Current Operating Procedures

- Spirituality NATI-style
- Nine pathways to truth
- Natural approaches to development

A General All-encompassing System of Comprehension

Due to the categorical nature of the NATI structure, we are able to place virtually any issue or problem into a particular category. We can then determine what the relevant data *behind* the issue is and then proceed to greater understanding or a solution. Since the NATI structure is based upon scientific principles of nature, it is a model greater than humankind, one that can be followed faithfully. This not only breeds confidence, but greater achievement since we can venture beyond our own limiting models, if we have the courage!

Moving beyond these limitations is crucial, however. I'm certain that everyone has experienced situations in which they felt lost, confused, uncomfortable, or out of place. It is not unusual for people at these times to fall below their normal levels of performance. In such circumstances, fear, emotion, and embarrassment are not uncommon. Not surprisingly, these situations are quite common. At the office, a cocktail party, or other social functions, we find ourselves blocked by the very walls we construct from our culture, our beliefs, our philosophy, our models of life, and more. We imagine something we want to strive for and see life through these elements, sometimes for better, sometimes for worse. What most people don't see is that they lock themselves into a mindset that, at many points, blocks them from integration, or being more intelligent and accomplished.

Development as Our Ongoing Mission

We are all familiar with the concept of building and thinking "out of the box." One of the biggest proponents of this concept is management guru Peter Drucker. Drucker states in a 1994 *Theory of Business* publication that, "The core mission of a company becomes obsolete from time to time. Unless someone is watching for new developments a company will find itself mired in the past. Without systematic purposeful abandonment of an old business theory, an organization will be taken by events.[1] Through Natural Intelligence Theory and, in particular, the notion of potential development, we can see beyond Drucker's theorem, for if we make development our core mission, issues such as "systematic and purposeful abandonment of policy" become extinct. This occurs simply because *development is a natural ongoing process of change, rethinking, and re-engineering.* The worst-case scenario is that a company becomes mired in development, which is not such a bad objective.

As quoted by a Dallas-based human resource firm, Sherwood Ross (a freelance writer who covers workplace issues for Reuters News Service) states, "the number one reason people leave or stay in a company is career development opportunities."[2] With the concept of "centering" development as the focus, everything and every event that follows has a purpose. Accordingly, when a company or group sets development as its main focus, the politics lessen, cultures become less restrictive, teamwork improves, and coordination (synchronicity) is enhanced. This all occurs naturally because all parties are participating in the same Point of Reference.

Organizing Data

Peter Drucker stated that information is the most important resource for executives. "Unless information is organized, it is still data."[3] You will recall in an earlier part of the book that we discussed the various stages of information and its acquisition (RIKU), data being one of the most basic components of how knowledge and information evolves. Drucker states that what makes communication in the workplace possible is focus, the best kind of focus centering on a common task or challenge. In NATI, the goal is development of the individual, specific sectors, groups, cultures, and, of course, the entire structure, whether it is a company, state, a nation, or the world.

In a general sense, Drucker confirms NATI's notion of multiple intelligences by stating that, "no two executives, in my experience, organize the same information the same way."[4] He further states that "information has to be organized the way the individual executives work, but there are some basic methodologies to organizing information."[5] One of them he calls "the key event."[6] He further goes on to state that "this key event may have to do with people and their development."[7] Indeed, this key event is, in NATI terms, an expression of development as the focus of the individual or the group. This notion is also echoed in a different fashion by Steven Covey, author of *Seven Habits of Highly Effective People*, through his notion of "principle-centered values."[8]

In *The Pursuit of Loneliness*, author Phillip Slater reminds us that the universe does not consist of unrelated particles but instead is an interconnected whole.[9] Pretending our actions are independent of one another is illusory. While our society gives more leeway for an individual to pursue his own ends, culture as a whole defines what is worthy and reliable. Balance is only achieved when most people pursue the same things in the same ways. However, the one thing that consistently demonstrates itself and is relatively simple to accept is the notion of development. If development becomes the "god of our children," humankind will be poised for an evolutionary leap well beyond that ever experienced by our species.

Business Tools as Proof

Being a businessman myself for many years, I can testify that NATI principles work in the corporate world with the same efficacy that they work in any other area of life. Indeed, many people have commented on the usefulness of NATI in tackling problems, and as part of this book's goal to emphasize real-life applications, it would be profitable to review some of these verbatim comments, all used with permission. The following comments were, in part, generated by executives such as Alan Johnson, former president of Macy's and the Bantus Group, as well as several other executives from smaller and mid-sized companies.

- Primary consideration regarding the approach, language, and presentation of NATI business tools . . . have to do with originality, power, scope, clarity, and evidence.

- Your non-judgmental approach is striking in that we know from therapy (both serious and pop) that a non-judgmental attitude is crucial in all kinds of transpersonal dealings, but the workplace is more commonly thought of as highly judgmental, so what you are emphasizing here fits in with what is known outside the workplace and needs to be brought into it.

- There's your emphasis on polarity. Various philosophers, from the pre-Socratics to the Renaissance on up to Bohr and Jung and Levi-Strauss in our own times have put great emphasis on polarity, and religion is catching up . . . I regard this as the area where NATI has the most to offer that's not available elsewhere in competing seminars . . .

- Scope is my term for the breadth of the [NATI] toolset. NATI proposes an inclusive toolset, one that can be applied to any problem because its scope is unlimited.

- What you are offered here is a philosophy of consciousness and it needs to be promoted and defended in the marketplace of philosophies, a marketplace which itself is currently split between the domains of philosophy proper, religion, and science.

In general, businessmen and women have found NATI to be a tool of both clarity and power, bringing insights to various strategies, problems, and decision-making by virtue of its principles and easy-to-understand vocabulary. It organizes data and makes the maximum use of available information, fulfilling Peter Drucker's belief, as stated above, that the "key event" for both people and business is development.

To achieve the greatest possible development, however, it is necessary to identify one's personal way of doing things, or what NATI calls Current Operating Procedures. We will focus on this now, as well as NATI's principles of wholeness, spirituality, wisdom, and pathways to truth.

A Wise Man's Notion of Development

The notion of development as the major underlying force of humankind and its purpose is not new. Aristotle discussed it at length. It has, however, been greatly underrated and misunderstood. This notion dates back to times of antiquity with such thinkers as Thales and Pythagoras, as well as great minds from the fifteenth to the twentieth centuries. In the 1500s, theologian and scientist Nicolas Cusa saw no limit to humanity's quest for knowledge and progress, stating "Is it not the end and aim of thought that generates unimaginable limits of convergent sequences, propagating itself, without end and ever higher?"[10] More recently, twentieth century anthropologist and theologian Teilhard de Chardin, a powerful force in Christian thinking, believed physical, spiritual, and mental evolution are one and the same process. "The role of man and the basis of all imperatives is to further that progress," de Chardin stated.[11] "Our existence is not a meaningless accident in an indifferent universe doomed to extinction, but the cutting edge of a process of universal evolution."[12]

Your COP

NATI defines the inherent intelligence already at work within us. It assists us in understanding and recognizing the tools we are already using—the thirteen intelligences. With Natural Thinking & Intelligence, we can better guide our behavior toward achieving a more satisfying and successful lifestyle. NATI is a development tool for enhancing our abilities to free ourselves from distorted thinking, to discover our potential, and to maximize our power. It enables us to more fully understand ourselves by making clear our thinking patterns and by helping us to recognize our "default" operating systems, or our Current Operating Procedures (COP). We repair our COP with a system we were meant to use before our behavior was corrupted by distorted thinking—institutional, fearful, and ego-based. Here are a few COP examples:

While researching and developing Natural Intelligence, I discovered that feelings of "superiority" can *at times* relate to striving for one's personal best. When people are moved to demonstrate superiority, they may be striving to simply show their perfect selves. Although not always the case, they are seeking the perfection stage of development, on a subconscious level. However, the notion of superiority can also occur when people try to overcome feelings of inferiority by repressing their actual feelings. This occurs through "superior" attitudes. They become very arrogant and tend to exaggerate their achievements. This is all part of an underlying pattern and model related to a distorted view of development within their subconscious. They are subconsciously seeking the state of perfection that we all strive for.

This brings us to the problems that "perfectionist" thinking produces, such as anorexia, bulimia, and other internal disorders, which stem from a lack of understanding. The process of life is not perfect. We never achieve a stage that is perfect no matter how good or elevated we may be. The best way to achieve perfection in life is to constantly pursue the notion of development. This then gives clear direction as to what decisions to make. Every person on the planet has an idea about what his perfect self would be like. Psychologist Alfred Adler called this "fictional fatalism." [13] Adler believed that it is impossible to totally understand a person without understanding the person's image of his perfect self. Once the unconscious understands this notion of perfectionism in the human psyche, it can then work in tandem with the conscious mind. Conflicts are then dramatically reduced and become much more manageable.

By understanding NATI, you learn to more fully develop those intelligences you are not currently utilizing as much as you could. You will heighten the ones you are already best at using, and you will begin to identify the restrictors that are holding you back: Fear, Ego, Ignorance, and Self-deception. We all allow one or more of these factors to hold us back to one degree or another. The more you begin to use your natural intelligence, the less emotional you will be. You will better evaluate decisions based on emotion and will use the very clear and natural powers of your mind. We will talk more about these in a later chapter.

The Natural Intelligence and Thinking System provides a key to successful living, which is accomplished through basing our Focus and Beliefs on development. Remember, your Character (and/or that which you are capable of becoming) is a product of your Awareness and Beliefs.

Defining Your COP

People solve problems and deal with life in their own ways, developing their own systems to deal with life. The NATI system provides you with an understanding of how you create, organize, and function by bringing you back to your natural self. If you want to redefine your life, you will need a new glossary. The qualities and virtues in nature mirror what exists within you naturally. Remember the "whole picture"—the connectivity within the universe that we have discussed.

Most of us have COPs that are based on comparing ourselves with others, but you will never be the best you can if you measure yourself this way. *Your only competition in life is reaching your goals, or your highest potential. You do not have to be like anyone else. You have your own special attributes that will take you wherever you need to go. Your level of accomplishment is up to you.*

Measuring your potential against your weaknesses, as opposed to someone else's strengths, is the first step. In order to do that, you must first understand what your strengths and weaknesses are. You must then learn to get in touch with what your true needs are. Your needs, quite simply, are your goals. When you learn how to identify your true goals, you achieve illumination. Once again, understand there are no limits on this journey except those posed by your own negative thinking. NATI principles are absolute and lead to infinite possibilities!

Recall our definition of potential: it is weakness unrealized.

We identify our mission, our purpose, and our level of potential by identifying and overcoming our weaknesses and turning them into strengths! At this juncture, we approach the realm of infinite possibilities.

The NATI Philosophy of Wholeness

Primary among the numerous tenets of NATI is that reality is harmonious and unified and that all things are interrelated. This is the concept of wholeness, which has a deep basis in both eastern and western thought. Unlike western thinking, however, eastern thought does not place an individual outside of what he thinks about. After all, how can anyone stand outside of all that exists? Amazingly, that is exactly what the NATI philosophy does. Have you ever noticed that as soon as people attempt to pin things down, they end up with only part of the picture? Ignoring or taking for granted the rest of reality, however, only limits thinking processes, and ultimately development.

Until now, it is the dominant belief that human consciousness can obtain only a vague sense of the entirety, or a whole concept, of nature. It is impossible to stand outside this wholeness and analyze it objectively. The NATI system, however, provides a format for doing just that. The methodology and degree to which we contemplate the whole is completely based upon the willingness, focus, and openness of the user.

NATI stresses natural wholeness and is based completely on the processes and structure of nature. According to NATI, the individual, nature, and society are actually manifestations of certain basic principles. As we shall see in the systems section, these principles are actually innate, not only to nature, but to each and every human being on the planet. Further, they are identified as intelligence.

Basic Distinctions

Eastern thought emphasizes less individualistic attitudes than western thinking. In the west, people tend to take pride in self-reliance and independence; in the east, the individual is considered as part of something greater. The NATI philosophy connects these two viewpoints quite efficiently.

One can argue that westerners recognize God as being far greater than the self, but that their God, paradoxically, has the overall effect of getting westerners to focus more on themselves. They see God as a personal being that cares about them and judges them as individuals. As a result, westerners often develop mental barriers that keep them separate from the rest of the world. This way of thinking encourages many westerners to be strong and enterprising, which admittedly is a positive outer demonstration of this type of thinking. However, it can lead to serious emotional problems, including loneliness, alienation, uncertainty, and feelings of guilt. In contrast, by not detaching the self from the rest of existence, NATI thought has a way of strengthening the emotional well being of those who follow it, thus reducing anxiety and isolation. In general, NATI thinking connects the individual to the rest of existence in three basic ways: by seeing the self as defined by its place in society, in nature, and in the universe. NATI teaches that by imposing a separation between the self and experience, one's experiences become less satisfying and spontaneous. Self-consciousness stands in the way of authentic experiences of reality in the fullest sense. NATI does not deny the individual; it simply says that we are interactive parts of a greater whole, and a vital one at that.

Another NATI finding is that we cannot rely on our feelings and impressions alone in order to give us a reliable sense of reality. We have to look beyond our individuality and personalities in order to recognize a greater dynamic. NATI provides us with the tools to do that if we so desire. Only by bypassing misguided, self-centered impulses can we free ourselves from what we identify in NATI as the Great Restrictors. Through this process, the mind is released and becomes truly free to be what it is. In NATI thought, the experience of freedom involves the merging of the individual consciousness with the universe or an absolute being (God) by recognizing ourselves as interdependent parts of a whole.

Spirit

In NATI, spirit relates to the inclination of potential. Because of the infinite and omnipotent nature of potential, there arises a sense of accomplishment, the realization of things that exist on a much higher and broader plane. Although we all actively participate and interact with potential, its scope is well beyond that of human reality. Human nature tends toward this sense of accomplishment since possibilities simply exist.

Most people often describe the difference between religion and philosophy in terms of devotional versus intellectual practice. These two kinds of activity have been separated for centuries, especially in the west due to the phenomenal success of western science. In NATI thought, however, the two are often complementary to one another. Spirituality and intellect are better able to work together in Natural Thinking & Intelligence because of NATI's tendency to avoid dogmatism, a form of closed belief. In NATI, spirituality or spirit is defined as energy rather than as a religious notion. This is a vital aspect of understanding human nature. It's a way of harmonizing personal feelings with groups, societies, and other cultures. In NATI, in fact, the term Directional Judgment means that a person's energy is focused on—or directed to—harmony, as well as abundance, acceptance, beauty, compassion, courage, truth, love, joy, and compassion. This energy, however, should never be confused with organized religious belief systems. NATI philosophy is non-dogmatic, rational, scientific, and intuitive.

NATI philosophy, therefore, is holistic, looking at reality as a whole in a constant state of flux. The purpose of NATI's tenets is to promote development in accordance with this unified whole (the Potential Field).

Distinctions and Beliefs

Science makes sharp distinctions between things and does little, if anything, to connect notions to an overall oneness. It tries to understand reality in order to control it, not to achieve harmony. Although the scientific worldview has led to the phenomenal triumph of Western technology and fostered an ideal of rational progress, it has left many in search of something more.

To its benefit, however, a distinguishing feature of science is that its objectivity and basis are verified by controlled, observable, replicable experiments. Anyone conducting the same experiment under controlled circumstances should obtain the same results. This enables all scientists to develop a common understanding of things regardless of what language they speak or what background they come from. NATI systems thinking achieves this as well. By experiential verification, one can be assured that scientific findings are independent of feeling, beliefs, and attitudes of those involved. This effectively eliminates wishful thinking, illusion, and hype. Consequently, knowledge, which comes through facts and scientific experimentation, doesn't depend on anyone's moral, religious, or emotional state of mind. Rather, it depends on the predictable, observable characteristic of things. This approach to knowledge-seeking has proved phenomenally productive in solving problems and conflict resolution. Who can question the dramatic improvement, due strictly to science, in the quality of life over the centuries? Science, lest we forget, has not only brought us enhanced technology and quality of life, but it has also contributed to the human condition of enlightenment, as well as personal and societal development in the realm of morality and emotional well-being.

NATI attempts to diminish harmful traditions, including corrupt concepts stemming from notions of nobility, elitism, position, and prominence.

A Perfect Model of Life

Consider the real estate agent who sells $350,000 condos. When a million dollar client walks in, or a developer looking to spend one to two million dollars, it is suddenly like a scene from "Lost in Space." He has no idea what he is doing. What happens to the real estate agent is that he is outside of his element. Put another way, no matter how successful that real estate agent is, and no matter how rich he becomes or how many awards he receives, he has adopted a behavioral mode, or mindset, that does not permit him to adapt to outside environments. The mindset cannot connect with anything larger.

This holds true for almost everyone. Most people limit the universe by their thinking. New experiences do not happen because their models are restricted. You may have heard the expression, "Humankind can't judge itself because it is the one doing the judging!"

Decoding potential is totally different. It is about a model of life that is not only outside of humanity but is perfect! This is so because it is self-organizing and carries an evolutionary focus. We are constantly surrounded by it every moment of our lives. No matter where or who we are, it is always there—effective, efficient, economic, and totally virtuous! At the same time, it is polarized, with both positive and negative elements present at the same time. Best of all, it is all-encompassing. Absolutely nothing escapes it, for it is nature.

Through decoding potential, we are attempting to bring to the reader an understanding of how nature works and its fundamental principles, as well as how to implement this model for any purpose. By using this model, one can learn to adapt, to achieve, and to develop!

Other Issues of Potential

If your situation is such that you are motivated toward making more money, gaining recognition or acceptance, achieving a promotion, landing that big deal, or getting somewhere with the "hottie" down the hall, you are in the mode potential. Decoding your potential will set forth the information and resources to assist you in achieving your goals.

On a deeper level, decoding potential will help you recognize your soul. It will enable you to understand yourself, your environment, and the events surrounding your life. Likewise, it will enable you to see into the souls of those you come into contact with in a totally different way.

Nine Ways to the Truth . . . at Least.

You might think that this book is going to give you the truth (or claim to know it). It won't. I don't know for sure what anyone's personal truth is. What I do know for a fact is how to *find* the truth. I do that through NATI, and so can you.

The reasons I am so emphatic about this are several. In the first place, twenty-five years of research and experience with NATI (and several thousand cases) represent significant proof of the program's effectiveness. Secondly, it has never failed. Finally, I have seen the reason why it works so clearly and definitively. It follows the principles of nature, not of humankind.

Let's briefly look at nine of the ways NATI shows us how to know the truth.

Mirroring–As mentioned in the organizational description, this is a very effective way of understanding. Mirroring accomplishes this end by showing us what we really are. In this respect, everything and everybody talks to us and is our messenger. They tell us what we don't want to hear but should listen to so we can understand our true inner selves. This visceral science is our inner spirit, our souls giving us notice and direction. As stated earlier, problems are our strongest common element. We constantly serve each other by virtue of bringing our problems to the forefront.

Objectivity–This is also tough. It is difficult to be honest with ourselves. In order to accomplish this, we need to look at events and behavior without opinion and then make commitments. The problem is our failure to be objective. Previous opinions, biases, and beliefs blind us. Failure to be objective disallows honest evaluation. In order to be objective, we need to be open to new data. (This reflects our definition of intelligence: the ability to recognize and integrate information.) Being blocked, therefore, inhibits our natural thinking and intelligence. The old joke that is applicable here is, "Don't confuse me with the facts; my mind is already made up." In order to be more objective, we should adopt perceptions instead of beliefs. This way we are much more open to adapt, to change, to see.

Impersonal–This is one of the most sobering aspects of finding truth. Particularly in business, one cannot take things personally. It generates emotion, and nothing lies to us more than emotions. The astronauts of Apollo 13 were totally impersonal when dealing with their life-threatening saga. Soldiers and pilots under fire survive because they are all business; they are impersonal. When emotion is taken out of the equation, clearer minds prevail. This is a key factor in attaining physical potential. The golfer, shooter, or tennis player who takes a "focus on the job" attitude versus a personalized, ego-centered outlook does not undergo the same stress factors.

Synthesis–This is a super process for finding the truth as well as expanding one's intelligence and understanding. We discussed this aspect to some degree in the organizational segment. As our definition of intelligence indicates, "the ability to integrate data/information into an overall meaning to pursue a concept" should take us to this point. When we arrive at the point of fitting a Concept/Belief into an overall picture, we see things, as well as ourselves, in another light. Sometimes we refuse to integrate, and it's at this point we have reached a moment of truth. We then need to make a decision to accept truth or deception (assuming we can find a way to fit the Concept/Belief into a reasonable whole picture). If one cannot do that, then outside assistance or open discussions with others may help one come to a satisfactory result. This entire process is great for dealing with the "forest for the trees" syndrome.

Weakness as Potential–In my work with clients, I am always amazed at how many people react negatively to this notion of weakness. More incredibly, it is successful people who reject this notion more so than others. I once had an attorney tell me, "I don't want to know my weaknesses. I'm afraid what I will discover." That's a rejection of truth by self-deception. Knowing who and what we are totally, both strengths and weaknesses, is critical to the truth, as well as clarity and understanding on a higher level. If we work on our strengths, we can achieve a certain degree of improvement. However, when we work on our weaknesses, the degree of improvement we experience is significantly enhanced.

Unconditional, Selflessness – For some, this is a very difficult notion to accept and follow. It models the "turn the other cheek" proverb. But consider this: we are always inclined toward taking care of our possessions and ourselves, putting ourselves first. That's perfectly fine! The problem arises when we lose sight of issues and become self-centered. We become biased and prejudiced against what is honest or fair. The lesson here is to take care of Number One, but only up to the point that it jeopardizes honesty and fairness. Being kind, fair, and truthful with ourselves and others is an invaluable quality.

Focus on Virtues–When I first began to realize the energy inherent within virtues, I was amazed. It became clear to me that focusing on a virtue and attempting to employ it opened myself up to unknown levels of inner strength. This strength gave me courage and motivation to pursue courses of action and, equally important, unlock an inner power and sense of well being. In 1981, it was this pursuit that led me to a state of total illumination. That's how powerful this process can be! Steven Covey relates this concept as being principle-centered in his book *Seven Principles of Highly Effective People*.[14] Every act represents a virtue or lack of one!

Development as Purpose–We have touched on this previously. However, in terms of finding truth and understanding, it is vital to grasp the fact that while having a purpose is good for us, developing potential is the key. This is because when we take responsibility for events and circumstances in our lives, we don't leave our fate to the heavens. While I believe in God, I find the notion of "placing my fate in God's hands" is simply an unproductive approach. In fact, I believe that we are *all* the personal manifestations of an impersonal force that I identify as God. I don't ask you to accept this. Just keep it in mind as you pursue your paths. The God force has given us the power to accomplish, achieve, and develop ourselves!

Following Proven Plans–In our Organizational Intelligence section we dealt with Laws and Models. This principle is akin to that intelligence. In this respect, if you need Models or Rules to follow in order to achieve clarity, truth, and understanding, this book is for you. Better yet, a good book on philosophy can set examples of proven roles, plans, or models. Of course, the best plan to pursue is following nature. That is precisely what the *Decoding Potential* books are all about, both Book I and II (and soon to come, Book III). The difference between following nature and following proven plans is a matter of scope and degree. There are more ways of achieving truth, clarity, and understanding than those cited above. In fact, the six organizational and three creative principles of NATI each contain inherent methodologies for attaining this end, some of which we have discussed earlier.

Normal versus Natural

One of the more interesting factors concerning the issue of decision-making is the natural approach versus a normal approach. Below is a chart whereby we can see at a glance the difference between our normal approaches to things versus a natural approach. With nature as our model, development of potential is always maximized.

NATI PRINCIPLES	NORMAL APPROACH	NATURAL APPROACH
Focus	What I want attracts me	Gathering info, seeking direction
Concepts	How does it fit into my mindset?	Establishing images for achieving
Expression	Doing what I want	Achieving purpose
Rules	I do what fits	Utilizing objective, successful models
Process	Rote	Systems order
Measure	Judging	Priorities
Mirror	What is the other person like?	What I am like?
Details	Looking at other things separately	Parts of a whole
Whole	I am the center	Integration
Physical	That which makes me feel good	Things to utilize for an end
Mental	Logic, rational	Systems thinking
Emotional	Satisfaction	Motivation
Spirit	Religious	Essence

Adopting a natural posture brings one into a greater state of mind, body and soul for the purpose of achieving development.

Chapter Eight

The Importance of Systems for Understanding

Systems are a part of our lives . . . and our lives are parts of systems, but we are scarcely aware of this fact. This is because we don't pay much attention to certain phrases we hear over and over again. We constantly live and interact within ecosystems, educational systems, weather systems, and systems of government. We use healthcare systems, communications systems, computers systems, banking systems, and transportation systems. These are just a few of the systems we experience on a regular basis, for all life on earth can be organized into social, technological, or biological systems. The point here is that we focus on the *type* of system we interact with, but not with the actual *structure* of a system. We take for granted assorted and questionable components of given systems, including the symbols used to communicate ideas, some of which are highly complex and sophisticated.

But what makes a system a "system? Is "system" just a name for a group of people or objects, or does the term imply something more specific, such as a network of relationships? It's important to answer these questions, for Natural Thinking & Intelligence is most definitely a system used for understanding.

Systems Theory

The theory of systems can be traced largely to one man: biologist Ludwig von Bertalanffy, born near Vienna in 1901. In the 1930s, von Bertalanffy formulated Organismic Systems Theory, an approach to systems in which a dynamic process exists within organic systems. The theory essentially focused on biological organisms in which metabolism strived for growth and equilibrium. Anyone the least bit familiar with high school biology will recognize that this describes cells, plants, animals, and humans.

In the 1940s, von Bertalanffy became interested in thermo-dynamics, or the transfer of heat into other forms of energy. Like his colleague Ilya Prigogine, von Bertalanffy was intrigued by the fact that the Second Law of Thermodynamics, which states that everything moves toward stasis and entropy, did not apply to all systems. Von Bertalanffy therefore formulated his Theory of Open Systems, showing that energy flowing into a system can bring it to a steady state (or a condition of self-regulation). If open systems did not exist, everything in the universe would soon run down, dissipating into chaos. Open systems are all around us, however, and import energy to maintain themselves: biological, meteorological, corporate, educational, governmental, scientific, and thousands more.

In the formal scientific study of systems in the 1940s, the terminology of biology and physics was used by von Bertalanffy and his colleagues. Von Bertalanffy wanted a more accessible and comprehensive approach to systems, however, and so introduced General Systems Theory. GST, as it is known, is a way of describing virtually any system in ordinary, non-formal language. Thanks to von Bertalanffy's simplified approach to systems, we can list the two most important principles of General Systems Theory:

- A closed system possesses components that interact only among themselves and not with the environment. This type of model is not a growing, evolving system. It is a system typified by rigid structure.

- An open system is comprised of components that can interact with each other after receiving an input of matter, energy, and information from the environment. According to von Bertalanffy, a system is not a "thing" so much as a dynamic pattern of organization in which the system as a whole is more important than its parts by virtue of the constant interaction of its components.

This is the essence of General Systems Theory, which has become a paradigm for understanding the dynamics of all social, technological, and biological systems since the 1950s. Part of von Bertalanffy's genius was to take a complicated set of variables and explain them in simple language that is universally applicable to all systems.

Open and Closed Systems . . . and Chaos

Systems that acknowledge their interdependence with their environment are open systems. Systems that don't are closed. This is the difference between being alive and being a machine. It's just that simple!

Machines are closed systems. They can only do what they were built to do. When a change in their environment occurs, they have no innate ability to adjust to it. They lose their relevance. When this happens, machines become obsolete and are consigned to museums or the scrapheap.

A chemical reaction that has reached stable equilibrium is a naturally occurring example of a closed system. Its principal elements are no longer exchanging electrons, absorbing or releasing energy in the process. The system becomes inert since energy has been evenly distributed. It has reached maximum entropy, the lack of energy available to do work.

Living things, on the other hand, are open systems. They adjust what they do and how they do it depending on the conditions they face, minute-by-minute and day-by-day, in order to optimize their chances of survival and well-being. We call them "adaptive" because they work to sustain their relevance and their connection with their environment. This continual interaction of components within a system causes tension and demands altered responses. Many people, especially traditional psychologists, view the absence of tension as the desired state of being. That's the human equivalent of the chemical reaction that has reached equilibrium or, in biological terms, homeostasis. It's a state of maximum entropy, for the system is inert or, in mental and emotional terms, either bored or dead. The tension caused by interaction with the environment *prevents* entropy, the loss of energy. It renews life. The challenge of creating new responses drives the system back to its core purpose so that it reinvents its forms and processes.

Human individuals (and social systems) can choose to be either open or closed. In an open social system, each principal component recognizes the other and the goal that unifies them. In a closed social system, each recognizes only itself, not the other or its unifying intention. This is the essential difference between actually being alive and pursuing a mechanical imitation of life. (The latter certainly describes most of humanity, which does not integrate healthy routines with enthusiasm, or potential energy, in order to enhance or improve them).

In closed systems, such as static institutions, the people in charge of maintaining the status quo force the rest of the organizational structure to operate in a robotic fashion. This produces entropy. Through it, these systems bring about their own demise. Some educational systems are certainly examples of closed systems since they are not open to evolving teaching strategies for fear of "rocking the boat". Schools themselves often cannot "learn how to learn."

In open systems, people get intrinsic satisfaction from continually interacting with the other principal components of the system. They use their individual creativity to solve present problems under present conditions. This perpetuates the system's well-being. A company will constantly reinvent itself in order to maintain a competitive edge. Apple, Inc. is a perfect example. Apple, makers of the Mac, was almost completely destroyed by Microsoft and its windows operating system. Macs, however, are now gaining in popularity because the company simplified its own operating system and made its PC more user-friendly. Creative energy literally reinvigorated the company and its products.

Here's a summary of the critical differences between open and closed systems.

Open System	Closed System
The system's originating purpose; the result it intends to produce.	Disregards original purpose; concerns itself with refining its forms and processes.
Connected; integrated with environment.	Disconnected; isolated from environment.
Dynamic; constantly reinvents forms or processes to sustain its purpose. This replenishes the system's energy and minimizes entropy.	Static; constantly strives for control and rigidity of processes. This prevents replenishing energy and maximizes entropy.
Integrative; synthesizing; sees wholes, their interdependent parts, and understands the relationship between them.	Linear and dissecting; analytical; sees parts in isolation, disconnected from one another.

(From: Meaning – Cliff Havenor)

What are Open and Closed Systems

Only things that have an inherent life force can create and develop new systems. Systems are of no concern to machines or rocks (unless one tunes into the Sci Fi Channel). Systems may originate as either closed or open. The difference between them lies in whether the system's originating purpose recognizes and unifies its principal components or whether the defined intent benefits only the system's creator. The former is an open approach. The latter is closed. For example, if a company introduces a product without describing (or even knowing) its full range of benefits, that represents a closed approach. If a company deliberately develops a product or service to provide what it knows will be distinctly beneficial to users that's an open approach.

The same is true of government. American democracy remains open to the extent that it constantly interprets the intent of the Constitution in light of new cultural patterns, technologies, demographics, political climates, and other factors. The Founding Fathers purposely designed a flexible document, not a rigid set of rules, since they understood that the world at large was ever-changing and not at all a closed system. Is democracy still an open system? I'll comment on that shortly.

Traditional evolutionary theory is a closed-system view of the creation of new species or systems. This theory contends that new systems are not deliberately created to be more relevant or better adapted to their environment. Those that prosper do indeed adapt very effectively, but this ability comes from accidents of gene recombination rather than from purposeful, knowledgeable design. This is "survival of the fittest," which, by the way, doesn't mean the biggest, strongest, or meanest son of a bitch in the valley always wins out. It refers to the success of those organisms most capable of integrating or fitting into their environment.

Examples of Closed and Open Systems

The average home heating unit is a closed system. Water is heated and sent throughout the house via a network of pipes. When a certain temperature is reached, the thermostat turns off the heater. It is an efficient system, but it is really nothing more than a feedback loop and is closed to outside information. This is true for all cybernetic systems, which are predictable because they regulate themselves within strict limits prescribed by their purposes and structures. Von Bertanlanffy believed that cybernetic models used circular causality and were therefore not dynamic. (Heated water affects room temperature, which affects the thermostat, which ultimately affects the heating of the water again.) He maintained that a system could be open only if it were non-mechanistic and were capable of handling the exchange of energy, matter, and information.

Open systems are indeed everywhere. Plant and animal cells are true open systems since they import energy in the form of chemicals, enzymes, or proteins and then utilize this energy to grow, multiply, and return waste products to the environment.

The extreme changes that can occur within a system are a demonstration of Chaos Theory. Chaotic conditions lie outside the normal limits of prediction or perceptual experience. (The odds of earth being hit by a comet are very slim, and yet scientists now know that such events happen at random intervals.) We will not go into the complex mathematical foundations upon which Chaos Theory rests, but what is important to understand at this point is that despite changes, a system can be restored to equilibrium.

NATI Thinking Systems

Let's take a look at how these open and closed systems work. First we need to be alert to the fact that there is a certain amount of abstract versus concrete thinking involved here. The chart below indicates the differences between abstract and concrete thought. It is a sampling Personality Orientation pairs, both abstract and concrete, and how they can be implemented into our own systems. Keep in mind that abstract notions tend to be open while concrete notions tend to be closed.

ABSTRACT	CONCRETE
spiritual	material
soul	body
theory	fact
design	structure
synthetic	analytical
intangible	tangible
holistic	fragmented
see	do
function	form
right-brained	left-brained
qualitative	quantitative
diverse	homogenous
distinctive	typical
inclusive	exclusive
people in charge of the system	system in charge of the people
creative learning	learning by authority

Any number of these Personality Orientations can be used to establish a variety of systems structures. When one considers the myriad of principles involved in decoding potential, it is staggering how many systems can be generated. Any one of these potential systems can be utilized for achieving a given result.

One major indicator of a concrete thinker is when a person attempts to qualify a principle. For example, a recent client came to me to define her NATI intelligences. We started with her awareness and her question was "awareness of what?" That's a concrete thinker. An abstract thinker would simply focus on awareness! A concrete thinker would say, "Why do I need to know the square root of sixteen? I'm going to be a tennis pro!" An abstract thinker would understand that the scientific principles and processes of square roots develop logic and a connective, synthesizing mindset.

Abstract thinking is important because it is absolutely vital for integrating things. Remember the essence of our definition of intelligence: "the ability to integrate information into a whole picture." It is also connected to open and closed systems in that abstract thinking requires openness, while concrete thinking is typically closed.

Creating Systems with NATI

At this point, we will first examine how we can analyze behavior and understanding with NATI. Using the example above ("awareness of what?), the client felt she was already a very aware person. We analyzed her awareness with the four functional principles (Physical, Mental, Emotional, Intuitive). As it turned out, she was very aware in Physical and Intuitive matters, but very weak in Mental and Emotional issues.

Moreover, we applied Awareness to the six organizational principles in order to further determine her *scope* of Awareness. Here's a summary of our findings:

	Organizational Principles	
Awareness of	Models – good	Abstract/open
Awareness of	Processes – fair	Concrete/closed
Awareness of	Judgment – good	Abstract/open
Awareness of	Mirror – poor	Concrete/closed
Awareness of	Details – very good	Concrete/open
Awareness of	Integration – fair	Concrete/closed

From the above, we can see that it's obvious that she was identified as a concrete, closed thinker with three-dimensional organizational awareness (four out of a possible six dimensions).

Let's go farther with this program. The next step was to analyze her belief, concept intelligence. We implemented the same procedure:

Concepts of	Physical principles – very good	Abstract/open
	Mental principles – fair	Concrete/closed
	Emotional – very good	Concrete/open
	Intuitive – good	Abstract/open

Here we found her to be a very functional person, with a balance of abstract and concrete thought patterns, and quite open. It is interesting to note that with this subject, her weakness (mental Concepts and Awareness) was her I.Q. This finding, as in many similar cases, emphasizes the value of NATI and questions our dependence upon I.Q. to provide an accurate picture of one's potential. Further, her planning/creative skills tended toward concrete (six out of ten of the above were concrete). Ironically, if this subject were able to convert her concrete classifications to abstract, *I believe her I.Q. would increase substantially!*

The Synthesis

The synthesis phase means recognizing the spiritual, emotional, and material states of a system, both its principal components and its original purpose. It doesn't mean throwing away what exists. It means discovering the meaning behind the system and a subsequent redesigning of the system based on its original intent.[1]

The synthesis phase is an open, adaptive system. It resolves the Catch-22s we see in the normal operating phase. People know the system's original purpose, or its "why." An integrative system is like water in its liquid state. It recognizes the four functional states and continually flows back and forth between them. Because it is tightly linked to the other principal components in its environment, it adapts its forms and processes to external changes. It is fluid rather than rigid.

If you understand the difference between the abstract and concrete phases of a system, understanding the synthesis phase is easy. Integrate the abstract with the concrete—the spiritual with the material—and you get a whole new system. This can be seen in thriving corporate structures that are essentially built on ideas. When ideas governing product design and development, management style and philosophy, and distribution of the final product are all seen as part of a single system, the holistic nature of the company can be readily seen.

People in open, integrative systems continue to acknowledge the system's origin, its principal components, and its intent. They understand the basis of unity between the principal components, even after the system has become large and materially complex. Therefore, they can see the meaning behind its forms and processes that result in the dynamic of maintenance and growth. They can see the relationships between causes and effects. They know *why* things do or don't make sense. They know what to change and when it needs to be changed. Any Wall Street investor who doesn't understand concepts such as cause, effect, or change is not going to be successful for very long, for the stock market is one of the most fluid models of a system that one can imagine.[2] Unlike operating within a typical system, whose complexity is incomprehensible, people can comfortably function in an integrative system because they have a foundation of purpose for organizing all the details. As organizationally complex as NASA is, its many employees are able to effectively organize their tasks so as to accomplish material goals (the building of space vehicles and millions of component parts), as well as spiritual goals (the exploration of space). Indeed, exploration is inherently an open concept, just as speculation and trading are open concepts in the market. Belief in a company represents an abstract (Belief), with a concrete (the company). The goods, services, and profits the company produces as a result of that belief are concrete.

In an integrative system (open or adaptive), people practice the inclusion of diversity rather than its exclusion. (The reader will recall that companies using cultural diversity can be highly successful.) They transcend dualism and understand the importance of using polarities for healthy functioning. This keeps the system integrated even after it has become extremely complex. They are concerned with both function and form because they focus on how things are complementary, how they "fit together." They remember that their goal is to accomplish the system's original intent.

We have many examples of open, adaptive, integrative systems in nature. Ant and bee colonies are two of the more obvious ones. In bee colonies, the queen, workers, and drones all focus on the intention of the hive, which is to sustain itself until a new queen is hatched so that other hives may be started. Any ecosystem, however, no matter how you define it, is a complex of interdependent, open, and adaptive systems. The intention of the system is always equilibrium capable of sustaining growth.

Within the realm of human experience, there are many examples of systems that *began* as open systems—systems of business, government, and education, but an integrative system *remains* open and adaptive after it is fully operational. By this criterion, the only examples of ongoing, integrative systems that I know of are specific people. We may say that integrative people:

1. Have a sense of purpose for their lives.
2. Are grounded in the originating purpose of whatever system they work in.
3. Are focused on the "outer/other" issue.
4. Seek the synthesis of separate issues. They are "unifieds," to coin a term.

Cross Discipline

So what is the resultant meaning of this Symbol and Archetype talk? Simply this: It's the ability to understand abstractly, but also how to integrate one system with another.

One of the many unique benefits of whole systems applications is known as "cross discipline." This is a process involving the utilization of the principles of one discipline and implementing them with another totally different one. For example, throughout Book I and II of *Decoding Potential* we have discussed General Systems Theory. GST involves utilizing principles of biology for human behavior. NATI utilizes multiple science principles for implementation with human understanding. The notion of hybrid might be applicable here such as cars which run on gasoline and electricity.

Chapter Nine

Integrating and Synthesizing Understanding

Polarities are everywhere. They exist within Derald Langham's geometrical model of the cell, and they exist in every human being. They are part of the cosmos and if we are to decode our potential we must come to grips with them.

Earlier, we established the importance of polarities and saw how they operate in various scenarios. Let's look at a few examples of integrating and synthesizing opposition from other areas of life before exploring how all of this information comes together in the dynamic open systems that are our individual characters.

Using NATI to Understand Systems

Just think about it: everything we see around us—all life and all social systems—evolved from simple one-celled organisms, which themselves evolved from strands of protein floating in the primordial oceans. According to the Second Law of Thermodynamics, there is no reason to believe that any complex structure should have developed, and yet here we are, discussing Natural Thinking & Intelligence. The movement toward growth and the actualization of potential is universal, and as I stated at the outset, this potential for development is written in the very structure of the cell as evidenced by Derald Langham's geometrical model. The thirteen intelligences that correspond to the thirteen stages of cell development also reflect the following concept: no single part of any system is more important to its functioning than another. In other words, the principles of NATI form a closed system and are fixed, possessing boundaries and precise structure, with all intelligences assuming equal roles. They are non-changing and invariant and relate to seeing things exactly as they are. Note that besides the thirteen intelligences, we can also include as absolutes both Polarity and Virtue.

How we *use* our intelligences is a different matter, for they are non-linear and capable of interaction in an endless number of ways. (An analogy is DNA, in which a limited number of fixed proteins can combine in endless ways, producing the diversity of all life on earth.) The conditions of our inner characteristics are variable and relate directly to perceptions as opposed to how things really are. In other words, NATI principles are *objective*, while our use of them is *subjective*. What this means is that Natural Thinking & Intelligence is composed of both open and closed systems as defined in the previous chapter. That is, Natural Intelligence is the closed system, while Natural Thinking is the open system. Our systems are always unbalanced, but by the same token, they are infinitely flexible, capable of restoring equilibrium to patterns thrown into chaos by the Great Restrictors (Fear, Ego, Ignorance, and Self-deception).

Before proceeding, let us look at the Human Character Formula again: A + B = C (Awareness plus Belief equals the

Character of our Communication) is a real example of a NATI system. Earlier we saw how the awareness of weaknesses relative to any number of personality orientations revealed our COPs, or Current Operating Procedures. What follows are case studies that include these principles and illustrate how important Belief is to a person's inner matrix, or life philosophy. In this study, we will see (a) how the absolute principles of NATI are affected by an individual's perception and (b) how the information plus the energy/tension of Polarity function in a system capable of realizing potential: the human being.

A Giant Gap

By 1992, new Head Coach Ray Handley had taken over the New York Giants from the very successful Bill Parcels, who had guided the team to a Superbowl victory. Handley's inability to motivate the team was implied by a headline in the *New York Daily News*: GUT INSTINCTS ARE CAUSING ULCERS. Handley was a highly intelligent man with keen Mental intelligence. In 1992, he attempted to teach his defensive unit, headed by all-stars Lawrence Taylor and Carl Banks, what amounted to a very cerebral defensive system. As the newspaper headline indicated, this intellectual approach did not go over well with the players.

The crux of the problem was that the players were all very much intuitively-oriented (such as linebacker Pepper Johnson), whereas Handley functioned Mentally and by-the-book. Not surprisingly, the Giants suffered a losing season that year. This is a classic case of how Mental and Intuitive functioning can conflict with one another. The results of Coach Handley's system were decidedly negative. While it may have had many redeeming features, Handley made no attempt to learn from the almost palpable team unrest. Weakness could not be turned into strength because there was no examination of the tension between polarities.

Handley was fired and replaced by the former head coach of the Denver Broncos, Dan Reeves. As the newspaper correctly noted at the time, good coaching is not possible without communication skills, and Reeves received high marks on this count because he knew how to connect with his players. Reeves taught them enhanced Physical intelligence, but he did it by helping players work on their fundamental skills, which was something to which they could easily relate. This truly represented a middle way. In fact, his approach was labeled "selectively aggressive." Far from being a contradiction in terms, this phrase demonstrated that Reeves was addressing the issue of polarity. He employed Mental tutoring as part of his overall system, but he did so in a way that increased the Physical and Intuitive capabilities of his players. With Reeves, it was not a case of either/or. He truly embraced the Principle of Complementarity.

Moving Toward Adulthood

In the mid-90s, Terry Anderson and I were doing some NATI programs together throughout the tri-state area. In Rye, New York, we did a seminar that clearly demonstrated the elegant simplicity of the NATI Potential Program to me, Terry, and other participants. About halfway through the program, we were discussing the practical applications of the NATI intelligences. We used everyday examples and proceeded to show how NATI can shed light upon circumstances and foster enhanced understanding. At one point, Terry brought up a personal issue to further explore the efficacy and efficiency of NATI. His thirteen-year-old daughter was, like most teenagers, anxious to grow up. She wished to start wearing make-up and was giving her father and mother a very difficult time about this issue. Terry's wife Maddie is a beautiful and intelligent woman with strong, fundamental, mid-Eastern Christian values. She did not want her daughter walking around looking like Bozo the Clown, as she put it. There was therefore conflict between mother and daughter. Having gone through something similar when my two lovely girls were growing up, I sympathized with her. We attempted to come to some sort of insight and direction regarding this matter utilizing the various intelligences, but without much success.

After about ten minutes, an elderly lady in the back of the room said, "The answer is Models." We all stopped, turned around, and looked at her. Terry asked her to expand on what she was saying. She repeated, "The answer is Models. What I mean is what you need to do is get somebody closer to her age—perhaps eighteen or nineteen—who will show her how to put on the minimal amount of makeup that will still be effective in achieving both ends of the spectrum to satisfy mother and daughter." Using someone closer in age, the old woman explained, would provide someone that Terry's daughter could better relate to.

We finished the discussion with Terry and his wife by selecting one of my own daughters as her model. Although slightly older than the model we had envisioned, my daughter had a strong connection with Terry's daughter and was able to successfully bring about a resolution that served all parties. This was a classic case in which Models was able to address Polarities: make-up/no make-up and young/old. Terry's daughter wanted to look older by using an amount of make-up that would probably have been inappropriate. (It certainly would have been to Maddie!) Finding a model who was young (but not too young) was the key.

I might add that, secondary to Models, Process was also involved since my daughter could explain how to be subtler in applying make-up while still appearing to be more grown-up.

Terry Anderson

In order to see the NATI intelligences working together, we shall look at two more extended examples that represent a more holistic model of the system. I want to begin with the ordeal of my friend Terry Anderson, the journalist who was abducted by Islamic militants in 1985 and held hostage for almost seven years in Beirut, Lebanon. Terry's plight was one that few people will ever have to face, but his survival shows that NATI is a dynamic system, one that can be used to solve even very serious problems by the application of the thirteen intelligences we have defined. Let's examine how each one was used by Terry during his captivity.

Focus/Awareness: Terry's *lack* of awareness was surely his downfall. He did not heed a kidnapping attempt the day before he was actually abducted, and he has stated in many interviews that, given the inflammatory nature of the Middle East, he was not as careful as he should have been. Once he became a hostage, however, his focus shifted from wants to needs, for survival most definitely did not include luxury items. He also accepted the fact that he might be confined for a long time and refocused his attention on his new, although quite undesirable, environment. Courses in survival training always teach people to focus on the immediate situation and tasks at hand. Panicking or continually asking "why me" not only squanders precious energy, but it also doesn't enable the brain to concentrate on the elements that require attention if one is to stay alive.

Beliefs/Concepts: Terry imagined himself to be a hostage, not a journalist. He accepted his fate and the fact that he was going to be living in a different culture. He knew the grim reality of his situation. He also realized that, in the short term, he was powerless to obtain his freedom. His mind came to grips with all of the unpleasantness associated with the concept of being a hostage. For Terry, assuming the identity of a hostage and believing in the reality of captivity was no different than someone believing himself to be a good citizen, educator, Muslim, Christian, or politician. It was who he was.

Communication/Expression: He therefore became a professional hostage and acted accordingly. His words and actions were commensurate with those of someone who is a captive. He could be very open or very silent, depending on what level of cooperation he wished to display. A wide range of social communication was no longer available to him on a regular basis, so he chose his modes of expression carefully. He needed to display both caution and presence of mind in communicating with his captors and fellow hostages. (After being released, he found his reintegration into society relatively easy because he had had almost seven years of practice in focusing his awareness on the manner of his expression. He used his recognition factor beneficially and communicated his ordeal in his bestselling book *Den of Lions*.)

Models/Laws: Although Terry followed the day-to-day rules prescribed by his captors, he also followed the laws of survival. His objective switched from the more immediate "I have to get out of here!" to "I am being detained against my will and I must stay alive at all costs." His new focus on the laws of survival was quite similar to that which people have when lost in the wilderness. Extreme necessity gives us many new Models that work with Priorities to show us what Procedures to implement.

Patterns/Processes: Terry adopted positive procedures. He asked for better conditions for himself and his fellow hostages as a way of implementing the laws of survival. At times, he absolutely insisted that certain critical needs be met. At other times, he was willing to negotiate. In yet other situations, he was completely passive since he realized that some conditions were unchangeable. In short, he did what needed to be done as opposed to what he wanted to do.

Mirroring/Feedback: As we have already seen in a previous chapter, Terry learned about his own behavior from the unpleasant exchange with another prisoner. He found that he himself was sometimes obstinate and inflexible (when such obstinacy was clearly uncalled for). He therefore learned from his experiences, both positive and negative. He was willing to experiment and try different methods of survival based on the feedback from both his captors and other hostages. It took a great deal of courage to look at his own shortcomings when the most obvious inclination was to focus only on the unfairness of his situation.

Assessment/Priorities: He decided what was important and how to achieve it. He established his dignity as a human being and stuck to it no matter what. When he wanted to communicate with his family, he continually battered his head against the wall— literally, even to the point of drawing blood until his captors relented. He paid close attention to the behavior of the militants so he would know when a straightforward request was likely to be honored, but he always assessed what it might take to effect change if those in charge proved intransigent on any given matter.

Parts/Details: Terry tried to make the most of everything, utilizing whatever resources were at his disposal. This is critical in any survival scenario. In a hostage situation, one makes use of simple things, such as the permission to move or speak, even if speech and movement are limited. A sip of water, a piece of bread—these are never disdained but are regarded as part of the larger picture of staying alive. Terry always tried to generate more resources, such as medicine, food, water, and exercise, as well as mental resources, such as the gratification that came from seeing that he could make himself heard. Perceiving success is an important aspect in the step-by-step process of staying alive in a precarious situation.

Wholeness/Synthesis: He brought everyone into the experience of being a hostage, even the guards. When he went into the bathroom and saw a machine gun leaning against the wall, he walked out and told the guard, smiling, that there was an unattended weapon in the bathroom. In short, he used physical, mental, emotional, and spiritual components to survive, and he worked with everyone he was with to achieve desired results. He learned that his captivity was a synthesis of east and west in terms of politics, goals, and behavior.

Lucy in Love

One of the most rewarding experiences I have had in assisting people with NATI came in the earlier years of its development, the fall of 1985. It involved a thirty-six-year-old woman named Lucy. With two small children (four and six years old), Lucy was a widow who had lost her husband Joseph within the past twelve months. Joseph had been a young man when he was killed in a tragic automobile accident. Lucy was referred to me by a mutual acquaintance who was a participant in NATI classes I was teaching at Seton College in Yonkers, New York. Apparently, she had been deeply in love with Joseph, and at the time of his death, they had been married for approximately ten years. By all accounts, it had been a fairy tale romance and marriage. She expressed the usual issues that are attached to this type of tragedy, namely grief, disillusionment, confusion, and loss. She could not understand why this happened to her and what she could do to help herself and her children.

Issues of this sort are never easily dealt with. They have a finality to them that is always sobering to those who are directly involved, and Lucy's case was just that. I spent an entire day with Lucy trying to answer her questions and generate a state of mind that would enable her to effect development for herself and her family. My greatest success in this area has always been in relating circumstances to absolutes such as virtue. By that, I mean that the circumstances involving people like Lucy and their spouses are actually representations of something on a deeper and more profound level. In Lucy's case, it was the virtue of love that she had experienced in her union with Joseph. I tried to explain to her that, in the cosmic scheme of things, Joseph represented the manifestation of the virtue of love in her life and it wasn't actually Joseph she was truly experiencing. It was the virtue of love. That is, Joseph was a concrete image, while love was the abstract. What I was trying to promote was that Joseph was an archetype of something that was very beautiful and prominent in Lucy's life. I know this sounds quite idealistic and abstract, and it certainly did to me at the time, but it was where NATI had led me.

After many hours, Lucy appeared to be somewhat improved in her emotions, but I could tell that she wasn't buying what I was selling. I suggested to her that the best thing she could do was to focus her belief system on the virtue of love and not on Joseph. I was careful not to denigrate her relationship with her husband, but rather to look at love as an abstract power source to help her get through it and find greater understanding and direction. She assured me that she would try this approach as best she could since it was really the only thing that made sense, although it lacked emotional satisfaction at the time. She left, and I felt good about the fact that I was at least able to give her some different insight.

I did not hear from her for nine months. One day, I was starting a lecture series at Seton College and, much to my surprise, I saw her listed on the attendance schedule. This was late summer of 1986. I started the class by simply introducing myself, and before I could get any farther, Lucy raised her hand from the back of the class and asked if she could say something. I naturally acknowledged her. She stood up and began to express her situation in detail, including the loss, grief, and suffering she had experienced. She also related our earlier meeting and how we had spent a good deal of time attempting to come to grips with her grief. She told the group that she'd had great disappointment and skepticism when we had concluded our original session simply because my words had not been what she expected or wanted to hear. She wanted her Joseph in the flesh! Nothing else had mattered.

Lucy went on to relate how she had searched for love in an abstract form and how things had started to happen in her life, such as meeting new people who were genuine, compassionate, understanding, and warm. She also stated how she began to see her life and the lives of her family in a different light. The notion of "falling in love with love" rather than with another human being was starting to take on a real life for her, so much so that she had begun to experience periods of elation. She concluded quite remarkably with the statement that the last six months had contained some of the most gratifying and compelling experiences she had ever encountered and that she was eternally grateful for my assistance and for the discovery of Natural Thinking & Intelligence.

For the next several months I couldn't help but think about the theory of Schrödinger's cat. Schrödinger's Cat Paradox refers to a cat being locked in a box with a device that may or may not release deadly cyanide gas depending on a single event: the radioactive discharge of an atom. As long as no one looks in the box, there is a fifty percent chance that the cat will be alive. Once the box is opened, only one possibility manifests itself. The cat is either dead or alive. Measurement is obtained by opening the box.

There was, and is, a question in quantum mechanics as to the measurement mechanisms of reality. One of those is known as linear measurement, which involves several possibilities of resolution. The other is known as relativity, which deals with a singular result. In the case of Schrödinger's cat, the measurement before the box was opened to see if the cat was alive or dead represented the linear measurement. There were any number of possibilities. Once the box was opened, however, other ways collapsed, and the only one that existed was the reality at that specific time. The cat was either alive or it wasn't.

This whole dichotomy of quantum measurement and its duality was, and always has always been, an issue. Thinking about what occurred with Lucy, as well as others I have dealt with concerning a focus on abstract virtue as a resolution, brought me to an understanding of a measurement process that reasonably addresses this quantum conundrum. I realized that the linear measurement was not an option since there was only one reality: Joseph was dead. The relative measure started to make sense when I realized that what happens when life concludes is merely a change of energy form, as in the case of the Second Law of Thermodynamics. There was a part of Joseph that took on the energy of the virtue of love. Lucy eventually related to that energy in a very strong way. What brought Lucy to the point of resolution was her search for the abstract energy form of Joseph, not the physical, material representation. She was able to connect with this energy and bring it into herself successfully.

I was able to get Lucy to regard Joseph as a manifestation of the virtue of love, which exists in timeless form. The resolution of her grief revolved around realizing that the infinite potential that we call love exists in a pure, perfect state. Being "in love with love" meant that she was able to regard the experience of being in love on an abstract level. For Lucy, love was no longer a single event but rather part of a much grander scheme that transcended a single person at a particular time.

Lucy is clear-cut proof that the notion of an abstract realm can have real applications if we are willing to entertain a paradigm shift in our lives.

Peak Performance Under Pressure

Clearly, one of the most valid accounts of peak performance under pressure came from golfer Ray Floyd. Floyd, an accomplished professional, was asked during an interview about performing under tremendous pressure. Floyd's response was classic in relationship to NATI's Patterns for Potential Program. In essence, what he stated was that pressure is a perception. How you view your undertaking has a direct effect on the amount and type of pressure produced. Further, Floyd adopted a very interesting mental posture. What he said was that he simply wanted to be in a position to be able to win on any given day. He didn't put pressure on himself to win every time. He considered himself no less of a golfer if he didn't win a tournament. His belief system was structured so that he felt he would be in a position to win within a given number of times, nothing more.

Pressure, then, is a self-induced phenomenon that is directly connected to Beliefs, Awareness, and Spirit. The Belief aspect is directly related to Confidence, while Awareness relates to Motivation and Focus. Spirit patterns can reduce pressure by focusing on Mission,

Task, and Purpose rather than on Ego or Fear.

The Human Character Formula is very relevant here, as always. Belief (or Confidence) is paramount. The ultimate Expression is changed by virtue of Awareness of a possible outcome in the context of a larger purpose: to play a good game of golf. With Ego negated, the Current Operating Procedure changes so that the inner matrix is healthier. While this process works, it is not easy to accomplish. However, with persistence it does indeed work.

In systems terms, pressure represents a tremendous amount of energy, but the energy can be redistributed so that the brain receives new information regarding the definition of success. Physical performance on the golf course then changes, showing once again that the interchange between energy, information, and matter can bring a system into equilibrium. The elements of Fear, Ego, Ignorance, and Self-deception, however, will keep a system closed every single time.

Olympic Pressure

Bill Mills, the 1964 Olympic gold medalist, had another perspective on pressure and performance. As a Sioux Indian, Mills was subjected to extensive discrimination for much of his life, which resulted in an indomitable attitude within the man. As he neared the finish line of his Olympic 10,000 meter race, he was in third place. His conscious thought was that he was going to be defeated, but he wasn't going to "lose." Third place in the Olympics is certainly no shame. However, when he accepted what seemed to be the inevitable fact of his defeat, he suddenly found new strength and sprinted dramatically into first place and won a gold medal. This perception—that one can be defeated but not be a loser—released large amounts of potential within Mills. In his Life Philosophy matrix, information was converted into matter and energy. Just as with Ray Floyd, Mills put himself in a position to achieve and then capitalized.

This concept is also demonstrated in the business world. Practically all functional people—the doers and the achievers—experience pressure from time to time since there are many factors that cause pressure. Pressure is categorically identified in NATI as an Expression. Expression, as we have stated earlier in the book, is the sum of Focus, Motivation, or Awareness, plus Beliefs or Perceptions. In this context, pressure is derived from the way people Believe, what they Believe in, their Interest, their Awareness, and their Motivation. It is also measurable in an Organizational sense, meaning that there are levels or degrees of pressure and peak performance. Further, pressure can manifest itself in any one of four ways (or combinations thereof), namely Materially, Emotionally, Mentally, and Intuitively. The same can be said to hold true for peak performance.

More NATI Validating

Terri Seppala, Senior Vice President and General Manager for Educational Services at CA GPS (Computer Associates International, Global Professional Services), was looking to change the way people are trained. What she had attempted to do was to establish a training/thinking tool that would map out a training schedule for each individual. Terri stated that, "We want to increase the speed of learning, along with the rate of technology change, in order to keep valuable employees."[1] She also stated that, "Students excel with different methods of training procedures, such as those that tap into their emotions, imaginations and other senses."[2] She believes that some of us learn better through different means and that we should be able to do much more individualized delivery. This is precisely the intent and direction of NATI. The thirteen intelligences are a powerful means to teach and understand how to realize our highest level of achievement.

The Laws of Elizabeth Smith

A number of years ago we conducted a seminar on decoding potential whereby individuals were asked to state the issues on which they wished to focus. At that time we were applying a very simple format to the program. We simply listed the thirteen intelligences in their respective groupings of Creative, Organizational, and Functional.

CREATIVE	Awareness
	Belief
	Expression
FUNCTIONAL	Physical
	Emotional
	Mental
	Intuitive
ORGANIZATIONAL	Rules/Laws
	Feedback
	Details
	Parts
	Order
	Assessment

We then put the stated issue up on a chalkboard and proceeded to relate the focus of the issue to each of the intelligences to see what we could glean from the exercise. At one point, an entrepreneur named Elizabeth Smith raised a business issue. Liz's issue dealt with a pending partnership arrangement with somebody she had been doing business with to some degree for the previous six months. We addressed her concern, which involved doubts she had relative to forming the partnership. She could not be more specific, but there were some underlying issues that needed to be uncovered. We quickly went through the creative aspects of Awareness, Beliefs, and Expression, but with little discovery.

We then examined the Organizational Group, and the first issue of Laws really set off the bell. Apparently, Liz's potential partner did not want to formalize the relationship with anything more than an informal memorandum. While Liz liked this person, she really didn't know if he was "the right person." Given the involved nature of the projects they would be working on, we quickly came to the conclusion that there needed to be some sort of formalized legal agreement put into place. Elizabeth assured us that this was going to occur.

Over a year later, I met her at another conference, and she explained to me that she and her partner had finally signed an agreement, although a major problem had arisen shortly thereafter. Because she had insisted on a strong agreement, things worked out in her favor despite some legal entanglements. She reiterated that if she had not pursued the course we uncovered at the NATI clinic, she might have ended up in considerable trouble.

This was a case in which someone simply needed to use her inner "lens" to focus on the intelligence of Laws. This alone was sufficient to prevent serious legal problems with her partner farther down the line. Paradoxically, she achieved her goal by *avoiding* the potential for trouble. Focus on Laws gave her the insight and courage to rid her inner matrix of self-deception regarding this man. Let me emphasize that this was done with a very simple format using a chart of the thirteen intelligences.

Synthesis

We can see from all of the above examples that we can use the principles of NATI to achieve development in so many diverse areas of life. What you decide to achieve is up to you, but it should be obvious by now that the thirteen intelligences are all interrelated and can produce outstanding results when we 1) use the Human Character Formula to focus our awareness properly, and 2) use the tension inherent in polarities to learn and grow.

The sky really *is* the limit! Every action is part of the mosaic of life. Using nature as our model, we can literally do anything, from becoming a better ice skater to overcoming grief.

Chapter Ten

Profiles for Understanding

A Circular World

In the mind/behavioral sciences of NATI, the world is circular, not linear. Instead of looking at data and events only in terms of logical, step-by-step processes, we need to view them from multiple points at the same time. The structure of these multiple points is actually nature and its principles.

Reviewing data in a logical order is quite limited for several reasons. First, if you don't know or don't have sufficient insight into a particular step in this process, you're stumped, delayed, or distorted. With circular or whole picture models, however, you view data and events by using any number of reference points or principles.

Secondly, with logic, a starting point is required. With the circular/whole model demonstrated in nature, you can start anywhere, just as you can trace a circle starting anywhere on its circumference. This is not the case when drawing a number line in algebra. A straight line has a beginning point and an ending point. It goes in one direction only.

Third, with logic you cannot move about various processes as freely as you can with circular, whole models. With the concept of circularity, you can go back and forth, up and down, and even expand to other systems in order to understand issues or solve problems. You can also incorporate negatives, weaknesses, restrictions, and polarities into whole systems, which is not the case with a logical, linear process.

A classic example of this is Einstein's formulation of the Theory of General Relativity. The heart of the concept occurred to him in a sudden leap of consciousness. As already noted, the proverbial light bulb went off while he was daydreaming (that is, thinking in abstract terms), not while he was hunched over his desk, feverishly scribbling equations with his pencil. With this leap, he was able to intuitively skip over the myriad mathematical hurdles of logic in order to reach his conclusion regarding relativity and then work backwards toward his model, sketching out the details after the fact, as it were. In essence, his epiphany occurred as a result of implementing a circular, "whole picture" model.

The major difference between the two models is that the NATI whole picture model is comprised of thirteen principles that, though connected, are individual systems. This model therefore generates infinite possibilities to realize potential, while logic, by and large, is limited. Many do not understand this because they search for the "how" and "why," looking for the logical sequence of an issue or event instead of seeking information that encompasses (or surrounds, as in a circle) what they are trying to grasp. But this is not how the whole systems model works.

Do You Get It?

A friend of mine read a rough draft of Book I of *Decoding Potential* and said, "I don't get it. Where are you coming from and where are you going with it?" I responded by saying that any of the book's principles, such as Nature, Complementarity, Polarity, potential development, the intelligences, and many more are part of a system greater than the human system. Accordingly, these key issues can embrace humankind, though the opposite is not true. As humans, we cannot totally fit these issues into our individual systems. Recall our earlier discussion in which we found that mankind fits within nature, not vice versa.

A friend and former colleague, Dr. Stephen Beller, is a licensed psychologist in New York. When I first introduced him to NATI approximately eight years ago, he just couldn't get it except for certain parts of the system. About six months later, he called me and excitedly proclaimed, "Bob, I got it!" He explained that, as a trained psychologist, he normally processed information by a system of logical classification. "It doesn't work like that," I told him. He had finally realized that by reversing the process and placing his own system within NATI, everything came together. He did indeed finally get it.

Dr. Myrna Watanabe, author and bioterrorism expert, expressed it differently: "The hardest part is letting go and putting your trust in nature. You have to adapt to it rather than try to incorporate it into your mindset. By doing so, your capabilities (potential) expand dramatically because you are utilizing a much greater system—nature!"

Understanding the Whole Picture Model: Formats

There are many possible formats we can use to further an understanding of NATI's approach. With these, you—the reader—will be able to get it, too! NATI's holistic model exists so as to enable everyone to have his or her own "leap" on the pathway to achieving potential. Below are some of NATI's key formats.

- COP: This is one's Current Operating Procedure. By virtue of historical data review, one can determine what procedures, patterns, or systems are currently in use. The factors we use to achieve this end are the same for all of the formats—the thirteen intelligences, plus Polarity.

- PIP: This is one's Personal Integrating Programming. It relates to how we integrate information and data into an overall picture or understanding, just as my friend Stephen Beller did. This is not a measurement device, but instead shows one how much work is required in this area.

- WL: This is one's Whole Life Profile. With this, one can examine the major and minor philosophical and conceptual issues of a subject. From this information, we can determine the basic factors that motivate and drive either beneficial or destructive patterns.

- AM: This is one's Achievement Model. With this format, one identifies the principles and intangible concepts that are at work under the surface of things. Every act we engage in or notion we entertain has some underlying concept. Once it is ascertained, enhancement almost always follows.

- PM: This is one's Priority Model. We think we know priorities, but experience shows us that our priorities are quite different than we think. They are also more fluid than we know. They are fluid, bending as the moment suits us.

- FM: This is one's Feedback Model. By analyzing feedback materials, we can easily see our direction. It mirrors a situation. This then identifies those issues capable of restricting our development.

- CO: This stands for Closed and Open Systems. While feedback gives us direction, it can also determine if we are open or closed to our environment. Hostility, bitterness, and rejection are some of the symptoms of a closed system (though not always). Growth, expansion, change, and intelligence denote openness. By applying polarity to the Human Character Formula, we can determine our CO.

THE NATI COMPOSITE

Primarily based upon:
A–Potential and its development
B–Polarity
C–Systems of nature
D–Problems/Weakness
E–Virtues

The Model is broken down into four categories:

NATI Principles	Scientific Basis	Mundane Applications*	Interchanging Opposing Positions
Models/Laws	GST, nature's structure	Values, opinions	Natural vs. Normal
Procedures	Systems	Patterns	Open vs. Closed
Mirror	Polarity	Feedback	Contrary vs. Harmonious
Measures	Relativity	Judgment, priority	Standards vs. Virtues
Parts	Multidimensionality, Continuum	Problems, issues, details	Parts vs. Entirety
Wholeness	Unification, Complementarity	Integration, adaptation	Entirety vs. the Parts
Focus	Development	Point of Reference	Notion of Development vs. typical, usual Perception
Beliefs	Potential	Frame of Reference	Cultural & Institutional Assumptions vs. Given Facts
Character	Potential development	Realized References	Realization vs. Unrealized

*How we utilize the NATI principles on a regular basis in our daily reality.

It's easy to see how applying these formats can help us better understand the big picture. Integrating information, being open, understanding priorities, using feedback, analyzing underlying concepts, understanding one's own life profile and operating procedures—these can all help us grasp more easily what is necessary for realizing our potential development.

NATI PROFILE WORKSHEETS
Worksheet #1

Determine your NATI strengths and weaknesses. Rate yourself on the following using a scale of 1 – 10. At the end of the exercise you should be able to detect where you need to concentrate your developments.

General Categories (Creative, Organizational, Functional)

		Sample answers*
creative/planning ability	_____	6
organizational ability	_____	3
functional/"take action" ability	_____	5

*14 out of a possible 30 showing a weakness in organization.

The Thirteen Specific Categories

		Sample answers*
focusing/awareness skills	_____	6
concepts	_____	7
communication/character	_____	3
adopting/following rules or models	_____	8
following given procedures	_____	6
prioritizing things	_____	2
implementing feedback	_____	3
integration of information/events	_____	2
adopting details	_____	1
physical action	_____	7
mental abilities	_____	7
emotional qualities	_____	3
intuitive abilities	_____	5

*In this example, the score is 60 out of a possible 130 with weakness in priority, integration, feedback, details. Strengths are concepts, models, physical and mental activity.

This exercise can be applied in general or to a specific situation or circumstance.

Worksheet #2

Write down an issue you wish to address. Make it short and to
the point.

_____

Then do the following:

Find your POR, or Point of Reference,
(your Focus)   _____

Find your FOR, or Frame of Reference
(your Beliefs)   _____

Next, assess each of the above as follows:

- Is your POR open or closed?   _____
- Is your FOR open or closed?   _____
- Is your POR objective or subjective? _____
- Is your FOR objective or subjective? _____
- Is your POR weak or strong?   _____
- Is your FOR weak or strong?   _____

From this exercise you should be able to view why you are getting the results you are (or why you are *not* getting the results
you are seeking).

For example: "I want to make more money" = my POR.

"I believe I can do it by making door knobs" = my FOR

Then:

- Making more money   = open POR – this is okay.
- By making door knobs   = closed FOR – uh oh!
- Making more money   = subjective – not too bad
- I could make $ making knobs = subjective
- Making money focus is strong = strong
- Doing it by making door knobs = strong
                                But it's not happening!

The problem lies in the closed FOR and subjective FOR = which
way to make $$.

Worksheet #3

Write down an issue you wish to address. Make it short and to the point. First, identify your Focus and Beliefs. Second, assess each according to the Core Human Dynamics. See which ones best explain your Focus and Beliefs.

_____

_____

_____

_____

_____

_____

_____

_____

Example:

"I want to make more money!"

My POR = making $$

My FOR = *need* to make more.

The CHD approach:		POR	FOR
Due to:	Want Control	NO	?
"	Power	?	Maybe
"	Attention	Maybe	Yes
"	Inclusion	Yes	Yes
"	Exclusivity	No	No
"	Measuring (my work)	?	Yes

This gives us some understanding about what's driving us toward a given objective. We can then take the steps, or paths, to achieve the same result.

Worksheet #4

Addressing the same issue, try to determine which of the components of virtue are missing. Depending on your answer, ask yourself what might work if implemented.

Note: The reader should read the first chapter before undertaking this task.

Example: "I want to make more money"

Courage   _____

Knowledge _____

Confidence _____

_____

_____

Let's also incorporate the qualities of nature set forth in chapter one for the same issue.

Effectiveness _____

Discipline   _____

Adoptive   _____

_____

Once we establish our weaknesses, our next step is then to strengthen them by whatever means available. For example, the three missing virtues may be overcome by hiring a competent, experienced manager to assist, or possibly a consultant or a partner.

The same can hold true for our qualities of nature.

Worksheet #5

Apply the same issue, and then answer the following questions:

What don't I know about this? _____

_____

Is there something I am afraid of? _____

Is my ego getting in the way here? _____

_____

Am I not being honest with myself about some issue? _____

_____

_____

In my previous example "I want to make more money":

What don't I know about this?
I don't know how to do this. I don't know how to set it up.

Is there something I am afraid of?
I'm afraid of losing my money.

Is my ego getting in the way here?
Not really.

Am I not being honest with myself about some issue?
I'm okay here.

The result here is the unknown factor. Also, what happens if I lose my investment? I'm not truly ready to risk the loss of money. Then it's not time yet. More research and background is necessary. Perhaps a financial partner or backer needs to be brought in.

Worksheet #6

Use the same issue, and then apply the nine steps for optimizing clarity and truth, comprehension, and your highest potential.

"I want to make more money"

Mirroring –	What is it that I like to do the least concerning this undertaking?

Objectivity -	Can I really do this materially, mentally, emotionally, and intuitively?

Impersonal -	Am I maintaining cool emotions and a clear head?

Synthesis -	Am I integrating the good points with the negative ones?

Weakness -	Look especially hard for weaknesses in yourself, the concept, the plan, and the organization. Then prepare yourself for having to deal with same.

Selflessness– Virtues -	Service to your customers See what works for you, the organization and your customers.
Development–	Remember to focus upon one objective: development and achievement
Proven Plan -	Look for successful models similar in scope and nature to what you are undertaking. Get all of the details you can and then implement them.

Any one or several of these exercises can enlighten you regarding any issue. After you have completed any number of them, put the information together and review the results.

Thinking in Thirteen Dimensions

There are certain qualities we all possess that make us what we are. As we all know, there are several methods available for identifying our make-up. At this point, let us look at another way of identifying our inner make-up. We call this "Thinking in Dimensions." Here's how it works. The participant answers a series of questions. The tally of yeas and nays will determine one's Understanding Quotient (UQ).

Here are the questions. Answer yes or no.

1. I don't take sides in issues such as male/female, religion, politics, etc. _____

2. I vote for the party; is only one true religion; the man or woman rules the the house. _____

3. I don't usually speak my opinion concerning a controversial issue. _____

4. Most of my social, casual conversations are about famous people, places, friends and events. _____

5. When rules, laws don't suit me, I follow my own. _____

6. My opinion is as good as somebody else's. _____

7. I do not accept criticism very well. _____

8. I'm part of the privileged, part of the select, etc. _____

9. I don't deal or associate with people I don't like _____

Yeas _____ Nays _____

Every Nay answer indicates a greater
Natural/System Understanding Quotient.

Natural Thinking Dimensions

Here's another model for thinking naturally vs. normally.
This is straightforward:

1. List the three NATI Groups and their respective principles.
2. Determine their applicability to yourself, or even better, to a specific situation.
3. Count the dimensions applicable.

For example: **The Creative Group**

I am a very focused person.	Y	N
I am very conceptual.	Y	N
I am quite a communicator.	Y	N

If you answered two out of three in the affirmative, then you would be a two- dimensional, creative thinker.

The Functional Group

I am a very physical/material person.	Y	N
I am very mental.	Y	N
I am quite emotional.	Y	N
I am very intuitive.	Y	N

If you answered three out of four affirmatively, you would be a three-dimensional action thinker.

The Organizational Group

I follow rules, laws, models (vs. opinions).	Y	N
I am a procedure person.	Y	N
I am a detail person	Y	N
I accept mirroring, critical feedback very well.	Y	N
I am not judgmental and take things relatively.	Y	N
I integrate things, events, people very well.	Y	N

So, if you got two out of six, you are a two-dimensional organizational thinker. In total, you would be a seven-dimensional thinker in respect to all three categories. The more dimensions the better since a greater number indicates an enhanced open systems orientation and a more significant orientation toward development

Case Number One:
The Case of Inadequate Tuition/Family Income

A client needed to make more money in order to support his family. Two children were headed for college within two years, and the family made too much to receive grants but not enough to meet his family's needs. He had three children, ages fourteen, seventeen, and eighteen. His wife worked as a secretary.

Analysis

Needs	= lacking	= negative
Make $	= create	= positive
More $	= growth, expansion	= positive
Too much income (to qualify)		= negative
Not enough $		= negative
Family needs	= lacking	= negative
Wife works	= produces	= positive

Results

There are three positives and four negatives. This indicates that there is a positive factor missing that will/can occur in order to offset the imbalance and create wholeness. In the negative aspects, there are two needs. Needs here are physical expressions. The other two negatives of "too much" and "not enough" are measures. The three positives were all creative. This indicates the absence of organization and action (function). This is ultimately reduced to "You need to better organize your actions!"

This eventually led us to reorganizing the family's spending habits, as well as coming up with tax savings measures.

What follows can be very effective. It's the Potential Intelligence Matrix (PI). It can be a bit confusing at first but stay with it because it works so well . . . with anything! It is also a classic utilization of Polarity and how to overcome it.

Case Number Two:
The Potential Intelligence Matrix (PI)

Here is an actual example of PI that I utilized for my books and seminars.

ISSUE – "I don't know how to promote NATI!"

First – I listed all the positives of NATI and its negatives.

	Positives	Negatives
A	13 Intelligences	1. Unknown
B	COF	2. Confusing
C	Development	3. No interest
D	Polarity	4. Insignificant value
E	Virtue	
F	Nature's Principles	
G	Life Philosophy	
H	Clarity/Understanding	

Next, I set off each against the other (Polarity) to see where they may coincide or offer possibilities. "Yes" meant it would not work, while "OK" meant it would.

	1	2	3	4
A	Yes	Yes	Yes	Yes
B	Yes	OK	Yes	Yes
C	OK	?	OK	OK
D	Yes	Yes	Yes	Yes
E	OK	OK	Yes	OK?
F	Yes	Yes	Yes	Yes
G	Yes	OK	OK	OK
H	OK	OK?	OK	OK

The summary indicates where the negatives might have an interest.

	Column 1	Column 2	Column 3	Column 4
Summary	Development & Virtue /Value Understanding	COF, Virtue Life Philosophy Positive Development, clarity	Development Life Philosophy Clarity	Development Virtue/Value Life Philosophy, Clarity

The analysis shows how many times each of the positive issues showed up across the matrix. The end result was that development and understanding was the path most likely to succeed since all of the negatives had an inclination toward those issues.

Analysis:

    4 – Development, understanding

    3 – Virtue, life philosophy

    2 – 0

    1 – COF

Chapter Eleven

Actual Models of Development and Understanding

Emerging Results

Throughout this book we have mentioned the notion of integration and development several times. At this point we will examine applications of these factors, and a few others as well.

Let's first examine the case for these key factors and their importance.

Dr. Ari Kiev, a clinical Associate Professor of Psychiatry, is the author of *A Strategy for Daily Living*. Dr. Kiev writes, "In my practice, I have found that helping people to develop personal goals has proven to be the most effective way to help them cope with problems. Observing the lives of people who have mastered adversity, I have noticed that they have established goals and sought with all their effort to achieve them. From the moment these people decided to concentrate all their energies on a specific, defined objective, they began to surmount the most seemingly difficult odds."[1]

Numerous other mind scientists like Dr. Kiev have the same opinion. And what better specific goal to pursue than *development as a life objective!*

A Scientific Doctrine of Development

Nobel prize-winning psychologist Ilya Prigogine won his medal for proving that certain impermanent forces, such as life forms, are able to produce open systems that both survive and advance in complexity. This fact exists even though these forms are closed systems! Something of a developmental nature is definitely occurring.

Additionally, the evidence presented by evolution shows that the entire universe is characterized by increasing order and recognizable structure. Accordingly, both of these scientific findings prove one thing for certain: the nature of the entire universe is gravitating toward development.

NATI People

Let's look at a few personalities that exemplify systems understanding. The first is a promising government official from Nassau County, New York, Tom Suozzi. Although Tom is not connected with NATI, he certainly exemplifies a systems thinker (and doer). It is therefore worthwhile to include his profile here.

Tom was elected as Nassau County executive in 1997. At that time, the county was rated as one of the worst run counties in the country, with an absolutely horrible bond rating. He ran against all odds and won because of his "can do" effectiveness. Within the ensuing eight years, Tom changed things around and introduced systems of governance that eventually saved taxpayer money and increased effectiveness of municipal operations. He integrated various departments to create new systems for saving money. For example, homeless people were not simply placed in shelters. They were individually questioned in order to discover what other private and public agencies might be of assistance to them. In some cases, it was the Veterans Administration; in other cases it was the Center for Alcoholism or ADD foundations. This thereby reduced the repetition of services and encouraged greater effectiveness.

Tom openly speaks of open and closed systems. For instance, he talks of how, in the past twelve years, there were more than 2000 elections in New York. Of that number, only thirty-four officials were not re-elected. This he calls a closed system and cites this as a basis for the stagnation and waste in New York government.

Suozzi uses systems to root out "do nothing" government workers and fire them, eliminate wasteful, repetitious contracts, and locate overlapping services. Tom Suozzi is a systems thinker who truly understands!

A Simple Game

Let's look at a very simple application of our A + B = C formula using another real life situation. This concerns a young Italian professional golfer, Marina Marselli. Until the summer of 2005, she had never qualified for a significant Ladies Professional Golf event. I saw her on the practice range one day and struck up a friendship. She was a good golfer with excellent mechanics. Late last summer, she was about to leave for a tour event in Canada. I decided to give her some assistance. The direction I gave to her was simply "Your center—your core—knows how to swing; so just focus on your center whenever you address the ball and swing! Let it go!"

The following week I stopped in the golf shop to pick up a club when she came up to me immediately. "Dr. Fiore (Flower in Italian), thank you so much!" she said as she hugged me. "I did what you suggested for the entire tournament. I came in second and shot the course record!"

Some of you may be thinking, "It can't be that simple." Well, here is another interesting observation: when our mind and spirit are open, the transference of understanding *can* be that simple!

Shoot Out

Here is another example. When competitively trapshooting in the mid-1990's, our five-man team was in a position to win first place. On the second round of shooting, our closest adversary shot very well. We were now under increased pressure and were quite "shook up". We had experienced this before, and what occurred was that negative energy took over and we began to miss targets. We were literally feeding on each other's negativity. This time, however, we were not going to allow that to happen. My colleagues, Sal Pepe, Andy La Salla and Steve Giamondo, decided to change our Model, our Process. What we did was split up the team and shoot separately in other squads. This not only dissipated the negativity—it actually enhanced our individual sense of responsibility and independence. Remember what football coaching legend Vince Lombardi said: "The crowd makes cowards of us all." So as not to feed off what we recognized as a negative situation or try to challenge that energy, we simply adopted a new paradigm. It worked and we prevailed.

I can relate numerous stories of this type in sports, business, and politics, where utilizing whole systems/NATI-principled strategies turned situations around dramatically.

Weakness as Potential

As stated earlier in both books I and II, weakness is our greatest pathway to developing our potential and learning how to understand. We have stated that working on our strength will develop us to some degree, but overcoming our weakness enhances our potential development immeasurably!

Here is still another interesting finding, and it's a beauty: If we follow the nature of the universe (and/or the universe of nature) and accept the fact that the universe, or nature, has no particular Frame of Reference, then weaknesses abate!

Are you still there?

Want an example? Here is a very simple, obvious one. If the Islamic Jihad did away with (or at least minimized) their cultural, personal orientation, there would be no Middle East conflict. Does that sound naïve or idiotic? Perhaps it does at this time in history, but the fact of the matter is this: it is another universal, infallible principle that works! If we worked on bridging gaps instead of blowing up bridges, maybe, just maybe . . .

How about another concept of weakness that actually was used to change our world? Some called it Peaceful Revolution. Two classic revolutionary leaders were Martin Luther King, of the Civil Rights Movement, and Mohandas Gandhi, of the New Republic of India. Both men believed in peaceful demonstrations as a way to create change even though they were being arrested, beaten, and humiliated.

The greatest problem with using weaknesses as a development factor is the unwillingness of people to face themselves. This is the downfall of people, society, and even entire nations. This is one of the greatest deterrents to understanding.

Personality Orientations (PO) and Understanding

POs are profile concepts, each of which has a polar opposite. Put another way, POs are Frames of Reference, which are limited, as well as Points of Reference with opposing profiles. Here is an example that demonstrates the importance of understanding POs.

Kids' Culture

Recently there was a headline in a New York newspaper that stated: "Teens follow styles, not rules."[2] This was very appropriate at the time since a rather prominent New York family with five children had contacted me to see if I could "do anything" with their adolescent kids and their growing misbehavior. You all know what it's like when kids hit that magic age (anywhere between thirteen and sixteen) when they abandon the nest: they know it all, can't be told anything, and will not do anything without their friends.

I decided to help them (even if I couldn't outright teach them NATI), but help is not "changing their minds." That's something no one can or should do. What these parents, as well as most parents everywhere, don't see is that, at this age, a cultural shift occurs. Kids adopt their own culture different from their parents. It is very difficult for kids to live in the family culture as well as a kids' culture at the same time, especially today!

What is needed is a transitional structure between both worlds. Kids and parents need to prepare themselves for this change in cultures. What both parties need to understand is that if adolescence goes "okay," it will eventually end. At that point the kids will come back to the roost. How they come back and how the parents receive them is another story. What generally occurs here is that this new kids' culture is an open system for them and a separation from the closed system of the family. What parents need to measure is the amount of organization that exists in the family unit prior to this change. This does not mean just discipline. It also entails order, virtue, and open-mindedness. The greater the structure of pre-adolescence, the greater the chance they will return to the roost "okay." This is experiential knowledge!

The culture we adopted for this family was as follows:

- Tell me all the good things about this new culture.
- Tell me all the bad things that occur within it.
- I will allow this (pink hair, punk clothes, rock music, etc.).
- I will not allow this (drugs, promiscuity, poor school habits).
- You, the child, must understand that breaking culture rules is one thing, but breaking civil rules is something out of the parents' control to fix. You are then on your own.
- If peer pressure becomes too great, and you (the child) become afraid of becoming an outcast, you promise to sit with Mom, Dad, an older sibling, aunt, uncle—someone within the family structure—to try to find a resolution to the perceived alienation.

In the case of the New York family, their fourteen-year-old girl was being pressured to smoke pot. She confided this to me and said she had already tried it but didn't want to hurt her family. We came up with a plan for her to advise her friends that she had an internal problem that could be exacerbated by pot. That took her off the hook but kept her in the circle.

China: An Emerging Systems Government

Stephen Johnson, senior policy analyst at the Heritage Foundation in Washington, D.C., was recently quoted regarding China's looming shadow in the world economy: "America's influence could be seriously eroded. The U. S. should rely more on competition in order to get a foot back in the door as a major player."[3]

When did we lose it? was my initial reaction. After some research, things became apparent, especially the differences between the two countries.

One of the big advantages is that China makes deals on the spot without strings, while the United States lumbers behind archaic, closed systems with loads of hierarchy. China has frequently sends high-level officials into small countries, treating them as members of the "big leagues." At the same time, American officials tend to ignore many of these smaller players, such as those from Latin America and Africa.

China is good for the developing regions of these countries. Investment in infrastructure and agriculture are neglected by the West but is much needed according to an article by Ambika Behal of the Washington UPI office.[4] According to Tom Friedman, the Chinese government is comprised of various types of professional engineers, while the U. S. is virtually exclusively attorneys.[5] The professional objectives of the two countries are like night and day, or should I say yin and yang. Basically, engineers are developers—open system types—while attorneys are not. They represent closed systems, are hierarchical, and are oriented toward the present moment. The U. S. needs to open both its systems and its eyes before the looming shadow of China and others, such as India, overpowers us.

Recently, China, a major user of soybeans, purchased the entire annual production of Uruguay's crop for the next ten years. Might this be a possible model for our oil consumption problem?

Here's What Brown Does

An up-to-date version of business systems thinking is exemplified in Thomas Friedman's best seller, "The World is Flat". Tom gives an example of "insourcing" which he identifies as a "whole new form of collaboration and creating value horizontally". In effect, he tells us, UPS has integrated within its operations a computer repair system for Toshiba directly with Toshiba customers! UPS has an entire facility set up for just this function. Additionally, they are dispatching the drivers and scheduling for Papa John's Pizza! Nike Sports also routes its good through UPS management as does Jockey.com.

As Friedman says, "This is not your father's UPS". This is cross disciplining, systems thinking in business–the parts contributing to the whole. He gives numerous other similar programs that companies are utilizing today.

A Systems Approach to Financing

During the earlier days of NATI (1980-81), the U.S. economy was in sad shape. At that time mortgages, if you could get one, were going for rates of eighteen to twenty-one percent! These were not good times for my industry, real estate. My professional background was in real estate and financial and investment consulting, so if I wanted to survive, it was necessary to come up with a vehicle that would enable my clients, partners, and myself to do business.

The answer was the Shared Appreciation Mortgage, which I developed in 1981. Briefly, the concept is structured so that a property owner/buyer would receive financing from a group/ institution as a mortgage with minimum interest payments. The owner would pay only an interest amount on the loan of approximately six to eight percent. They would give up a portion of the property's ownership to the lender. The lender would receive, at some time in the future, a portion of the increase in the value of the property. For example:

- Property value $100,000
- Loan $70,000
- Base interest payment of $5,000 per year
- Ownership interests: 50% owner, 50% banker/investor

- Property value increase over a two-year period = $25,000 +
- Investor/Banker's return: $12,500
- Investor/Banker's percentage of return (per year): 16% + ($22,500 ÷ 70,000)

When you have no alternatives, this is a great deal for both! This is how I implemented NATI in the discovery of this concept.

The focus:

- Buyers and property owners needed affordable funding.
- Bankers and investors required security and higher than normal returns.

The concept:

- For buyers and property owners willing to pay but who couldn't afford monthly payments
- For bankers and investors willing to wait for the return on their investment
- A + B = C

By integrating the "can do" with the "can't do," I came up with the notion of avoiding higher than normal monthly payments in exchange for means to accomplish a house purchase. What is interesting to note is that by the shared appreciation method, the owners would pay less than if they took out a straight eighteen percent loan. Therefore, for the home purchaser, paying $12,500 to the investor in increased property value (plus $10,000 in payments) equaled $22,500 versus eighteen percent interest payments on a $70,000 mortgage for two years (or $25,200).

This concept was first written up in the Stock Market Magazine in June, 1981, and later in the *Wall Street Journal* and the *New York Times* in February and April, 1982 respectively.

How to Be God . . . Like

In the Book of Proverbs, there is a saying (25:2) that states: "It is the glory of God to conceal a matter; to search out a matter is the glory of kings."

I believe that there is only one God and that what we regard as the quality of "all knowing" comes from this one God. Such a "knowing" includes all the religions of the world, as well as all individuals and organizations that do not believe in the existence of a God at all. Call this God a Force of Nature, the Universal Mind, the Great Spirit, the Great She—whatever you wish. It is only a matter of semantics. This force is the knowable essence from which all things come. *All* religions are true, and each is but another form of striving to know and develop to a higher form of humanity, one that transcends our present condition and rises to infinite and total well-being and oneness with the universe.

Abraham, Krishna, Buddha, Zoroaster, the Devil, and Jesus Christ are all personifications of this God force. All men and women on this earth, regardless of race, nationality, or creed, are one big family, one people, and stem from one universal consciousness.

The Science of the God Potential

How curious it is that the definition of potential identifies so well with a definition of God. The accepted definition of potential is "something that does not exist, but can be imagined or conceived, something that is latent. Further, when we look at the qualities of potential, we see the very nature of God, for in all religions, God is conceived to be the sum of all things possible."[6]

From this universal definition of potential we can glean that it must also include our three absolute components: Planning (Creative), Organizing, and Functioning. We know this is the case because reality can only fall into one (or a combination) of the three. It is also a fact that these three can occur in a positive or negative manner. We now have a description of potential that states that it is an "Omni force that exists within each of us that does not materially actualize without us."

To reiterate, this Omni force has three parts to it: the Creative, the Organizational, and the Functional. It also has a positive and negative aspect to the three parts in that it contains both good and evil, or Polarities.

Earlier, we defined intelligence as "The ability to recognize data and integrate it into a whole picture." Ultimately, this is the ability to contain a totally embracing frame of reference, with that entirety being the *main* point of reference. Not only is this the alpha and the omega point, it is the singularity—the pre-Big Bang state of transcendence.

Finally, understanding development as a vital factor is directly connected to many religions and societal concepts, which are closed, while development itself is an open concept. The difference between these factors and nature is that nature is understandable, scientific, and realistic. Even when we encounter the mysteries of nature, they are acceptable because they are real. Development is real, and no God, personal or impersonal, would ever condemn developing potential.

Chapter Twelve

The Genius Within

Up to now we have discussed a number of notions concerning achieving potential, development, principles of nature, NATI's thirteen intelligences, problems, Great Restrictors, and more. Let's examine what we have thus far.

At the outset, we are all much better and more capable than we act! As stated earlier, it makes no difference what your IQ is versus your UQ (Understanding Quotient) or your PQ (Potential Quotient). To further define these quotients:

- UQ – One's ability to see through things, to get to the heart of the matter; to put things (such as information) together into a comprehensive picture; to expand one's Point of Reference and Frame of Reference without foregoing one's ethics; the implementation and utilization of data and knowledge into meaningful concepts; to see the end before events begin.

- PQ – The Physical, Mental, Emotional, and Spiritual levels one can achieve; the willingness to strive for higher levels or goals; seeking to discover paths for expanding abilities.

As you may agree, neither your Understanding Quotient (UQ) nor Potential Quotient (PQ) requires a significant IQ. Instead, they require one to alter his or her Point of Reference and Frame of Reference. Moreover, they require a positive, or at least a neutral, approach to the Core Dynamics mentioned earlier (control, power, etc).

How Do We Do It?

Being natural (as opposed to being normal) is critical if we are to enhance our understanding and our lives. In the last twenty-five years of NATI research, I have become very disturbed as I've witnessed how the world treats itself. We accept kids into the school system who are slow, have weak memory skills, and have minimum maturity levels. We then tell them their skills are nil and turn them into minimum wage hot dog vendors, never to be heard from again. At the other extreme, consider bright, alert students who can't gain acceptance to an Ivy League university and settle on a small Midwestern college. Their chances for acceptance to an Ivy League brokerage firm (or other prestigious company) is virtually non-existent. By the same token, we have heard Donald Trump talk of his disappointment in MENSA apprentices. The point is obvious. These kinds of limitations are without merit in most cases and are therefore totally unnecessary. In fact, it's possible to take uninformed, limited individuals and turn them into accomplished human beings by showing them how to actualize their potential.

Instead of citing copious examples, let's look at the underlying factors—the NATI intelligences—to demonstrate.

First, let's consider the Focus/Awareness intelligence. In practical terms, this is what we identify as one's Point of Reference. Instead of the mindset of "I'm limited," we focus on strengths and weaknesses. We make weaknesses an object of potential. Adopting a model and process is the best way to achieve this. The individual's judgment is switched from "I'm unable" to "How much progress am I making?"

Second, let's examine the Concepts/Beliefs intelligence. This is related to one's Frame of Reference. People tend to formulate notions, pictures, and belief systems of themselves and their lives. They attempt to filter all events and data through their particular Frames of Reference. In so doing, they eliminate, or at the very least, limit achievement. The reason why some people accomplish so much is because their Frames of Reference enable them to do so!

One of the methods we use to enhance our understanding and achieve greater potential (through Beliefs/Concepts) is to reframe our Frames of Reference to include missions and objectives, such as development. Some of the more concrete objectives are accountability and responsibility. Achieving these objectives takes victimization out of the equation. Once we eliminate the belief that we are victims, we have no excuses to limit ourselves or to practice self- deception. These are just some of the details required in order to achieve integration within the new Frame of Reference.

The problem people have with understanding is that they are not open to personal critique. That means their Points of Reference and/or their Frames of Reference are too closed. Earlier we discussed open and closed systems in human dynamics. This is where these systems are especially applicable. If our POR and FOR are closed, there is no way we can integrate or synthesize data or events in our lives. There is no integration or understanding! We may gain knowledge perhaps, but not necessarily understanding. Closed POR and FOR automatically eliminate events and data that don't match up with their mindsets. Remember open systems grow and expand, while closed systems eventually die.

God's POR and FOR

Let's put this entire scenario into a God perspective. Simply put, God has the ability to totally integrate anything. God is wholeness and oneness without separation. Everything is included within God. If this is true, and most people tend to accept it as so, then God can't possibly have any preferred Frame or Point of Reference! But how about good and evil? These notions change nothing! Good and evil can exist within wholeness or oneness and still not be active. It is the human condition that activates them.

The A + B = C Genius: Communicating our Reference

The third aspect of our Human Character Formula, A + B = C, relates to our Communication or Expression. This is important for two reasons. First, Communication is a very important intelligence. People who communicate clearly, distinctly, and fluidly are very effective. One might identify this quality as the "glib tongue." We have all heard the expression "He can talk his way out of anything." Effective communication can therefore be a very important and efficient principle to practice in the right circumstance.

The second aspect is the utilization of communication as a basis for problem solving and conflict resolution. This is achieved by the systems approach of the A + B = C formula. Whatever is expressed or communicated is a result of the Awareness or Focus (our "A" factor) plus whatever we believe about that focus (our "B" factor).

Another way of looking at this is: Whatever a communication refers to is the result of a Point of Reference (Focus) plus a Frame of Reference (Belief)! We do this automatically everyday in everything we do, but we have no understanding of this process—until now!

Functional References

At this point, we utilize the four functional intelligences—Physical, Mental, Emotional, and Intuition—by integrating them with our Communication/Expression references. These are the only ways we can express or act!

Accordingly, when we talk about our Human Character Formula of A + B = C, we can now begin to see things in a much clearer and more concise fashion. For one thing, when we are seeking conflict resolution or problem solving, we can recognize if the issue is Physical, Mental, Emotional or Spiritual. This is the reason why some consulting or therapy programs work and others don't. Some may not be addressing the proper functional principle! For example, motivators address Emotional focus, but the issue may be Physically based. The Emotional approach may work temporarily but will eventually dissipate.

Additionally, functional principles and the A + B = C formula demonstrate how people can confuse or restrict themselves. Take, for instance, a situation where one's Awareness is Spiritually inclined while his or her Belief is Physically oriented. The resulting expression may well manifest itself as emotional conflict. The following scenario may apply:

- A Spiritual focus – It is blessed to give to the poor.
- A Physical belief – I enjoy material possessions.
- An Emotional expression – I feel guilty about buying those new shoes.

This kind of inner conflict happens to us every day in a myriad of ways.

As you can see, the potential for the clear, holistic framing of one's reality is immensely important within this system!

The Genius of Polarity

Let's look at how the A + B = C principles are impacted by Polarity, or opposition. The first polarity we shall address here is open versus closed systems. Confusion and chaos often occur when one's Point of Reference and Frame of Reference are at odds, such as when the POR is open, but the FOR is closed. For instance, a man focuses on making money—an open POR—but believes he can only do it through the manufacturing industry—a closed FOR. He has statistically and realistically reduced his potential by closing himself to two of the biggest proven ways of making money: real estate and the stock market.

In over twenty-five years of working with NATI and whole systems practice, I've learned that a closed system is the biggest restrictive factor of all. This finding holds greater validity with the FOR.

A closed POR may be easily altered. For instance, if one's POR is on a democratic politician, all one needs to do to change the democratic POR is to start talking about the benefits of a two-party system. However, the FOR is something else! Here we are talking about Belief. One of the best sayings that relates to Beliefs is: "Don't confuse me with facts; my mind is already made up!" The more closed the FOR, the less the chance for growth, change, or achievement.

There are certainly times when a closed FOR is beneficial, but that is when token mechanical and restrictive aspects are involved. An example of this involves U. S. Open golf champion Retief

Goosen of South Africa. Goosen is one of the top five golfers in the world today. He has a unique system he uses to overcome the tremendous pressure of the PGA tour. When he needs to come up with a critical shot, he "concentrates on concentration." How NATI is that? Here, a double FOR and POR utilizes a double abstract closed system in order to achieve an open system result!

Surgery is another example of a closed system. No one really wants his doctor "poking around" during surgery. Concentration and Awareness are crucial.

How Organizational Principles Can Bring Out Our Genius

As you can imagine, pursuing thinking of this nature can uncover some very interesting findings. As we have said before, systems thinking is the combining of various principles or disciplines, thereby creating new ones. In effect, we integrate. Accordingly, we can then take our six organizational principles and apply the aspects we have been discussing. For instance, we can apply open and closed aspects, as well as Polarity (oppositions), to each of the six in order to create a new approach to a given situation. If we wanted to investigate whether or not our organizational principles are open or closed, we would create a simple chart as follows:

	Models	Process	Priority	Feedback	Detail	Integrate
Closed						
Open						

At this point I would ask the following questions:

- Am I (or another person) open or closed to utilizing a new model?
- Am I open or closed to feedback?

To some this may seem very simplistic, but I can assure you I have taught this system to athletes, business people, and professionals who have implemented it to the fullest.

Sometimes people utilize the program for one thing, and another end result manifests in a totally different area. Someone many readers may relate to concerning this is golf teaching pro Jim McLean. I knew Jim as a young assistant pro at Westchester Country Club in Rye, New York back in the late 1970s and early 1980s. Jim has always had an ebullient personality. We played golf together, and I took lessons from him. In the early 1980s, while NATI was in its infant stages, I began to use self-hypnosis to get individuals to access their natural talents. At the time, Jim was interested in developing his golf skills further. Jim began to work with me at my home, where we had discussions about teaching styles. I was always of the opinion that the best teachers adopted a method to fit the student. I honestly don't know if Jim felt that way before our getting together, but obviously when one looks at his teaching career over the last twenty-five years, one can see how diverse his approaches to golf are. During our sessions, Jim was very intense in his application, quite open, and committed to the priority of development. While Jim did not become a noted touring pro, he has become one of the top teaching pros in the United States, being named Master Teacher and Teacher of the Year in 2000 by *Golf Digest*.

Whole Systems for Organizing

When we look at our organizational principles through whole systems thinking, we see a totally different view of things. Here are some views created by implementing the organizational principles, together with some Cultural Characteristics.

- Are the models we use open or closed?

- Is the process we implement flexible or inflexible?
- Are our priorities weak or strong?
- Do we analyze feedback objectively or subjectively?
- Do we examine details positively or negatively?
- De we look at the whole picture with an exclusive or inclusive perspective?

These are just some of the methods we can use to gain greater understanding through organization. Therefore, let's look a little deeper at NATI's ability to generate whole systems. As we have seen, this is achieved by integrating various NATI concepts and principles, thereby creating new paths to understanding. The chart below is just one approach to "whole system creating" with the organizing principles.

ORGANIZATIONAL WHOLE SYSTEM CHART

	A + B = C	CHDs, CCs, etc.	POLARITY	INTEGRATION	MIRROR	PRIORITY
LAWS, MODELS	The Human Character Formula	Results of our Concepts-CC's always paired	We all possess good and bad, which is normal	Synthesize everything; put it all together	Polar opposite observations	Is always development.
PROCESS	Focus plus Expression	Impacts our habits/patterns	Define the opposition	Seek what both sides can accept	What we don't like in others is within us	As always, how to develop
MEASURE	Result of A + B	Underlying influences	Based on the CHDs	Try to synthesize	See ourselves, then decide	Always based on development
FEEDBACK/ MIRROR	Expression = A + B	Directs us toward our underlying dynamic	Feeds us the underlying core components	Continuous alteration of data toward a whole picture	Shows weaknesses	Treat as development insights
DETAIL	Every conceivable Concept	The specific CHDs, GT, Restrictors, etc.	CHDs, Omnis, GT. Restrictors, and CCs	Individual parts, but part of a greater whole.	What to look for in the mirror	Each part is separate but part of something greater
WHOLE	A + B = C Potential and Polarity	Parts of the Whole	Polarity is part of Complementarism	A Whole System, a true intelligence.	Clarifies the entirety	Integrate thesis into a comprehensive schema

Virtues: True Inner genius

We have spoken of how using the thirteen intelligences in a synergistic fashion is analogous to a symphony, where notes or instruments are blended to produce a synthesis of various parts. It is not an exaggeration to say there will never be a point at which composers will scratch their heads and say, "Oh well, it's all been done! There are no more symphonies to write." If possibilities are endless for composing a musical score with a finite number of notes (think of a piano keyboard—it is long, but there are only so many keys to work with), just imagine the possibilities that exist for the human brain to generate ideas using its trillions of cells.

The history of mankind's evolution indicates that there may well be no limit to potential. If there is an endpoint to mankind's overall development, however, it may be similar to Teilhard de Chardin's Omega Point, where human consciousness will be a single entity—not a population of individuals. Even if this is the case, one may ask the question: What then? Will the collective consciousness of humankind be faced one day with new lands of choices, such as the exploration of different dimensions? Indeed, will this new entity be in a position to *create* new dimensions or universes? Will the finished product of evolution, as many mystics and scholars have said for centuries, be part of what we now call "the Godhead"? These are tantalizing concepts depicting an abstract, transcendental, and yet scientific existence.

Actually, the good news is that we already have access to absolutes that can help us transcend the limits of our everyday lives to one degree or another. Yes, the NATI intelligences are absolutes, but what I am referring to is virtue, pure and simple. I am not talking about Sunday school mentality here, nor are you, the reader, in peril of receiving a sermon. As you know, NATI is completely non-doctrinal, and I suggest that virtues are irreducible in nature and can, in and of themselves, help to unlock a person's potential.

Virtues Over Values

You realize the best results in achieving your goals when you follow patterns of nature, patterns that naturally lead to development. Using concepts of potential development as the prime motivators for your very existence—the explicate manifestation that is you—can literally change your life. The same holds true for focusing on values and virtues and then adhering to them. As Socrates said, "Virtue is its own reward." This is why I firmly believe that when you attain virtue, you have already begun to realize potential in your life. The reasoning behind this is that virtue is a pure, clear path for pursuits.

I am, of course, making a fundamental distinction between values and virtues, for what one values in life is not necessarily a virtue. Simply consider people like Hitler or mass murderers like Ted Bundy. Values can be terribly skewed by the Great Restrictors we have alluded to so many times. The restrictors can corrupt our values and the matrix in which they exist. Values, however, tend to be less concrete than virtues, which is why adopting solid values is extremely important in one's struggle toward development of values that reflect virtues. In his book *Seven Habits of Highly Effective People,* author Stephen Covey says that it is essential to be "principle centered."[1] It is extremely beneficial to find important principles and values that you can believe in and then bring them into your everyday life. In effect, they can become the "seat of your soul." In NATI terminology, this is the basis of your matrix.

Virtue, as we said, is another matter since virtue is absolute and is intrinsically related to Plato's concept of Ideal Forms. A virtue can never be reduced any farther than its given definition or corrupted, as values can be.

Examples of virtue are abundance, acceptance, patience, tolerance, endurance, balance, consideration, clarity, courage, strength, fortitude, honor, honesty, truth, integrity, caring, cooperation, love, beauty, elegance, refinement, and confidence. I consider all of these to be NATI points of excellence that define a superior path to achievement, change, and growth. Significant power is mobilized in one's matrix when these virtues are implemented. In earlier chapters we chronicled many contemporary ills—problems in education, government, and religion—in showing the universal need for NATI. Consider for just one moment how various institutions—closed systems, if you will—could be changed for the better with the power of virtue as a driving force behind man's creativity. Individually, our COPS would be changed because Core Human Dynamics would be used in a more positive manner. Power and control would be used to accept people and their ideas. Drive and motivation would be aimed at the common good. Attention-seeking would not lead to ego, but rather to a healthy self-image and a feeling of uniqueness. Value judgments would be fair and unbiased, with creative expressions stemming from a focus on (and belief in) virtue!

If you think all of this is too idealistic, you might be right. However, ideals are attributes that can direct our matrices away from restrictive notions and provide a clear direction. Consider some of the advantages of pursuing virtue:

- They give strength. They are positively oriented, with no negative feedback, either consciously or unconsciously.

- Inefficient patterns are put into perspective, enabling one to discard distractions and focus on goals.

- Virtues flow naturally, reducing stress since focus on them distracts the mind from tension.

- They are self-sustaining. They provide confidence and motivation in public and personal life because they cannot be jaded or transcended. They are strengths in themselves and have no agenda.

- Virtues are self-organized. They innately lead to better intuition, organization, and functionality. They continually interconnect in a positive manner.
- They are always practical. They are easy to follow because they are identifiable. Following them becomes second nature.
- They produce expanded awareness and can see all parts of a situation. They rise above duality (that is, the either/or mentality and quantum measurement).

Some may believe this is all very impractical, but today pragmatism is very corruptible and is, in many cases (if not most) the easy way out.

Practicality of Virtue

At this point a practical example is in order. Not too long ago, I was taking a golf lesson from my friend John Kennedy of Westchester Country Club in Rye, New York. John and I are always discussing ways golfers can achieve a greater degree of potential. I was in a terrible slump and couldn't break out. During the lesson, John asked me how this book was coming along and what were some of the keys. At that point, when I mentioned virtues and their application, something struck me. I was not applying the notion of virtue during this golf slump. I began to question John with the objective of reducing the mechanics of the golf swing down to what virtues may be at the root cause of my poor swing execution. It didn't take long to discover that the two primary *missing* virtues were *patience* and *trust*. For you golfers, I was not "waiting" on my swing, and I wasn't trusting it either. After one round of golf focusing on these two virtues, I was almost totally back to a better game. Business and professional applications are no different! *I have come to see that failure, mediocrity (and other problem areas) are due to a lack of some virtue.*

Virtues as Systems

We discussed systems at great length in earlier chapters. Virtue can enhance virtually any system—personal, societal, or cultural. Notice how many of the characteristics mentioned in the section above relate to General Systems Theory! Virtues, when incorporated into *any* system or matrix, behave like dynamic components. They flow naturally and are self-sustaining and self-organized. They are also balanced, the very heart and soul of Complementarity. They constantly give strength, supplying energy, a vital component to any system. All of this is to say that virtues are not subject to any laws pertaining to dissipative structures, such as the Second Law of Thermodynamics. The reason for this is that they are absolute. They cannot be broken down or changed. They are incorruptible. This is why I emphasize their potential and transformative powers.

The question, therefore, is not so much whether or not we should adopt virtues. That is a given. Rather, what we should ask ourselves is: "Can we afford to live without them?"

And Finally . . .

The word "natural" is prevalent in today's society. We want foods that are natural and free from harmful chemicals, preservatives, and pesticides. We want herbal remedies, aerosols free of CFCs, asbestos-free and lead-free building materials, natural exercises (such as yoga), natural diets, cleaner fossil fuels, solar energy. . . and the list goes on. In short, we want things that work in harmony with nature in order to be healthy and live long lives. But aren't these goals? Of course they are. We have a natural tendency, conscious or not, that wants to realize potential. And this is the revelation of NATI: there is a natural intelligence *within* us, a natural way to think, and there is no activity in the entire world that we can perform without thinking. If we crave things that are natural, then we need look no farther than our own thought processes.

The Thrill of Victory and the Lessons of Defeat

Although the decoding of potential has set forth scientific and philosophical means for touching deeper levels within ourselves, the development does not always come easily. Achieving one's potential is akin to the undertakings of an Olympic athlete. The pains of Olympic training are very much in league with what some of us need to go through in order to achieve desired results. Focus and belief are not just words. They are calls to action, so to speak. Moreover, defeat and failure are also part of the game. We learn by experience, to some degree, but only if we accept the negative feedback of failure and prepare ourselves for its reoccurrences.

Commitment to Development

The commitment to the development of potential is far and away the most important responsibility of all humankind. Nature has given us all the natural abilities we need. However, this not the end of the story, for such a book as this can have no end without thinking on some level. You, the reader, are the logical conclusion to these chapters . . . and you have just begun.

Summary

While values can be skewed by the Great Restrictors which can corrupt our inner matrices, virtue is absolute. With virtue, our adverse COPs can be changed since Core Human Dynamics can be used in a more positive manner. Virtues give strength, put inefficient patterns into perspective, flow naturally, reduce stress, are self-sustaining and self-organized, are practical, and expand awareness. In short, they have transformative powers. Don't be afraid to use them! Likewise, don't be afraid to seek out and use your innate, natural intelligences; they are your soul!

Book III to Follow

| The Natural Intelligence Group ||
Natural Intelligence	Natural Thinking
INVARIANT PRINCIPLES Relates directly to seeing things as they are	**VARIANT PRINCIPLES** Relates directly to viewing things as one sees
Polarity 1. Positive (+) 2. Negative (−) 3. Neutral (0) Direction	**The Great Restrictors** (What Blocks Our Achievement) 1. Fear and Anger 2. Self-Deception 3. Ignorance/Confusion/Doubt 4. Ego
The Human Character Formula A Awareness [A] (Information) + B Beliefs [B] (Knowledge) = C Character [C] (Understanding)	**Core Human Dynamics** (Our Natural Motivation) 1. Power 2. Control 3. Acceptance/Inclusion 4. Uniqueness/Exclusiveness/ One-Upmanship 5. Attention/Recognition-Seeking 6. Self-Interest 7. Judgment/Values/Appraising 8. Motivation/Drive/Will
The Four Human Functions 1. Physically 2. Mentally 3. Emotionally 4. Spiritually/Intuitive/Instinctively	
The Six Ways Humans Organize 1. Models, Laws, Rules 2. Procedures, Methods, Processes 3. Measures, make Judgments and Appraisals, Priority 4. Give and Receive Feedback 5. Segregate, Divide into Parts 6. Integrate, Synchronize, Combine	**Steps for optimizing Comprehension, Clarity, and Reaching our Potential** 1. Objectivity, Openness, Flexibility 2. Follow a Proven Plan 3. Be Impersonal 4. Unconditionality, Selflessness 5. Use the Principle of Development as Resolution and Life Mission 6. Synthesize and Integrate evens and Information. 7. Use Weakness as a Potential 8. Use the Mirror/Truth Seeking 9. Stay focused on Values and Priorities

NATI PROFILE WORKSHEETS

Worksheet #1

Rate yourself on the following using a scale of 1–10 for strengths OR 1–10 for weaknesses:

Creative/Planning ability   _____

Organizational ability   _____

Functional/Take action ability   _____

Rate yourself on the following utilizing the same scale above:

Focusing/Awareness skills   _____

Concepts   _____

Communication/Character   _____

Adopting/following rules on models   _____

Following given procedures   _____

Prioritizing things   _____

Implementing Feedback   _____

Integration of Information/Events   _____

Adopting Details   _____

Physical Action   _____

Mental Abilities   _____

Emotional Qualities   _____

Intuitive Abilities   _____

Worksheet #2

Write down an issue you wish to address. Make it short and to the point.

Then:

Find your Point of Reference, your Focus _____

Find your, your Beliefs _____

Next:

Assess each of the above as follows:

• Is your Point of Reference open or closed? _____

• Is your Frame of Reference open or closed? _____

• Is your Point of Reference objective or subjective? _____

• Is your Frame of Reference objective or subjective? _____

• Is your Point of Reference weak or strong? _____

• Is your Frame of Reference weak or strong? _____

Worksheet #3

Write down an issue you wish to address. Make it short and to the point.

Identify your Focus and Beliefs. _____

_____

_____

Then assess each by the Core Human Dynamics (your natural motivation). See which ones best connect as an explanation of your Focus and Belief.

Power _____

Control _____

Acceptance/Inclusion _____

Uniqueness/Exclusiveness/
    One-Uppmanship _____

Attention/Recognition-seeking _____

Self-interest _____

Judgment/Values/Appraising _____

Motivation/Drive/Will _____

Worksheet #4

Addressing the same issue, try to determine which of the components of virtue are missing–what might work.

Examples of virtue are abundance, acceptance, patience, tolerance, endurance, balance, consideration, clarity, courage, strength, fortitude, honor, honesty, truth, integrity, caring, cooperation, love, beauty, elegance, refinement, and confidence.

_____

_____

_____

_____

_____

_____

_____

_____

_____

_____

_____

_____

_____

Worksheet #5

Apply the same issue; then address the following questions:

What don't I know about this? _____

_____

_____

_____

Is there something I am afraid of? _____

_____

_____

_____

Is my ego getting in the way here? _____

_____

_____

_____

Am I kidding myself about something? _____

_____

_____

_____

Worksheet #6

Apply the same issue; apply the 9 Steps for optimizing clarity, comprehension and your highest potential.

Objectivity, Openness,
Flexibility _____

Follow a Proven Plan _____

Be Impersonal _____

Unconditionality, Selflessness _____

Use the Principle of Development
as Resolution and
a Life Mission _____

Synthesize and Integrate
Events and Information _____

Use Weaknesses as a Potential _____

Use the Mirror/Truth Seeking _____

Stay Focused on
Values and Priorities _____

Any one or several of these exercises can enlighten you regarding any issue. After you have completed any number of them, put the information together and review the results.

References
Chapter One

1 H. R. Pagels, *The Cosmic Code* (New York: Simon & Schuster, 1982).

2 *Dictionary of Science,* New York: McGraw-Hill, 9th Edition, 2002.

3 Gregory Bateson, *Mind & Nature.*(New York: Dutton Publishers, 1979).

4 David Bohm, *Wholeness & The Implicate Order* (New Jersey: Prentice Hall, 1980).

5 P. Russell, *The Global Brain.* (New York: St. Martins Press, 1983).

6 J.D. Barrow, *Theories of Everything* (New York. Fawcett, 1991).

7 Fritjof Capra., *The Tao of Physics* (New York: Bantam, 1976).

8 J.S. O'Connor and Ian McDermott, *The Art of Systems Thinking* (San Francisco, CA: Thorson's Publ. 1997)

Chapter Two

1 *Oxford Companion to Philosophy* (London, England: Oxford University Press, 1995).

Chapter Three

1 Ann Weiser Cornell, *The Power of Focusing* (New York: MJF Books, 1996)

2 Ibid.

3 Terry Anderson, *Den of Lions* (New York: Crown Publishing Group, Inc., 1991).

4 Rupert Sheldrake, *Seven Experiments That Could Change the World* (Los Angeles, CA: Berkeley Publ. Group, 1995).

Chapter Four

1 Howard K. Bloom, *The Lucifer Principle* (Berkeley, CA: Publishers Group West, 1995).

2 Abraham, Maslow, *The Farther Reaches of Human Nature, 2nd ed.* (New York: Viking Press, 1971).

Chapter Five

1 William Blake, quoted in *Major British Poets of the Romantic Period*, ed. by William Heath (New York: Macmillan, 1973).

Chapter Six

1 Diane Dreher, *The Tao of Inner Peace* (New York: Harper Perennial, 1991).

2 L. LeShan, *The Mystic, The Median and The Physicis* (New York: Viking, 1974).

3 J. Needham, *Science and Civilization in China* (Cambridge, England. Cambridge University Press, 1956).

Chapter Seven

1 Peter F. Drucker, *Innovation and Entrepreneurship: Practice and Principles* (New York: Harper & Row Publishers, 1985).

2 Ibid.

3 Ibid.

4 Ibid.

5 Ibid.

6 Ibid.

7 Steven Covey, *Seven Habits of Highly Effective People* (New York: Simon & Schuster, 1990).

8 Philip Slater, *The Pursuit of Loneliness* (Boston, MA: Houghton Mifflin, 1970.

9 *Oxford Companion to Philosophy* (London: Oxford University Press, 1995).

10 P. Russell, *The Global Brain.* (New York: St. Martins Press, 1983).

11 Ibid.

12 Abraham. Maslow, *Towards a Psychology of Being* (New York: Viking Press, 1962).

13 Steven Covey, *Seven Habits of Highly Effective People.*

Chapter Eight

1 Cliff Havener, *MEANING: The Secret of Being Alive* (Edina, MN: Beaver's Pond Press, 1999).

2 Ibid.

Chapter Nine

1 Highbeam Research.com http://www.highbeam.com/terriseppala

2 Ibid.

Chapter Eleven

1 Dr. Ari Kiev, *A Strategy for Daily Living* (New York: Simon & Schuster, 1997).

2 USA Today. Pg. 6.

3 Highbeam Research.com http://www.highbeam.com

4 Ibid.

5 Ibid.

6 Tom Friedman, *The World is Flat* (New York: Farrar Straus & Giroux, 2005).

7 Gregory Bateson, *Mind &* Nature (New York: Dutton Publishers, 1979).

Chapter Twelve

1 *Steven Covey, Seven Habits of Highly Effective People* (New York: Simon & Schuster, 1990).

BIBLIOGRAPHY

Anderson, Terry. *Den of Lions*. (New York: Crown Publishing Group, Inc., 1993).

Aquinas, T. *Summa Theologica in Selected Political Writings*, Ed. A.P. d'Entreves. Oxford:University Press, NY and London, 1948.

Ardrey, R. African Genesis. New York: Dell, 1961.

Asimov, Issac. *Science, Numbers & I*. Garden City, NY: Doubleday, 1968.

Bale, Lawrence S. and Gregory Bateson. *Cybernetics and the Social/ Behavioral Sciences*. www.narberthpa.com/Bale/lsbale_dop/ cybernet.htm.

Banting, P.M. "Marketing, Scientific Progress, and Scientific Method." *Journal of Marketing* (1978), 99-100.

Barrow, J.D. *Theories of Everything*. New York. Fawcett, 1991.

Bateson, Gregory. *Mind & Nature*. New York: Dutton Publishers, 1979.

Bentov, I. *Stalking the Wild Pendulum*. New York: Dutton Publishers, 1977.

Beukema, P.L. *Predicting Organizational Effectiveness with a Multivariate Model of Organic and Mechanistic Value Orientation*. Unpublished doctoral dissertation, University of Southern California, Los Angeles, CA, 1974.

Bloom, H. K. *The Lucifer Principle*. Berkeley, CA: Publishers Group West, 1995.

Bohlen, J.M., C.M. Coughenour, H.F. Lionberger, E.O. Moe, and E.M. Rogers. "Adopters of New Farm Ideas: Characteristics and Communication Behavior." In *Perspectives in Consumer Behavior*, Eds. H.H. Kassarjian and T.S. Robertson. Glenview, IL: Scott Foresman, 1968.

Bohm, David. *Causality and Change in Modern Physics*. Philadelphia: University of Pennsylvania Press, 1971.

———. *The Enfolding-Unfolding Universe.* Interview by Renee Weber in Revision, Summer/Fall, 1978.

———. *Quantum Theory.* Englewood Cliffs, NJ: Prentice Hall, 1951.

———. *The Physicist and the Mystic—Is a Dialogue Between Them Possible?* Interview by Renee Weber in Revision, Spring, 1981.

———. *The Special Theory of Relativity.* New York: W.A. Benjamin, 1965.

———. *Wholeness & the Implicate Order.* New York: Prentice Hall, 1980.

———. and F. David Peat. *Science, Order, and Creativity,* London: Ark, 1987.

Boulding, K.E. *Evolutionary Economics.* Beverly Hills, CA: Sage Publications, 1981.

Bradshaw, Gary. *Wilbur and Orville Wright.* www.wam.umd.edu/~stwright/WrBr/Wrights.html.

Briggs, John P. and F. David Peat. *The Looking Glass Universe.* New York: Simon & Schuster, 1984.

Brown, R. "A Brief Account of Microscopical Observations." *Philosophical Magazine, 4,* 2000, 161.

Buckley, P. and F. David Peat. *A Question of Physics: Conversations in Physics and Biology.* London: Routledge & Kegan Paul, 1979.

Burr, H.S. *Blueprint for Immortality: The Electric Patterns of Life.* London: Neville Spearman, 1972.

Cairns, Huntington. *Legal Philosophy from Plato to Hegel.* Baltimore, MD & London: Johns Hopkins Press, 1949.

Calder, Nigel. *The Key to the Universe.* London: Penguin Books, 1981.

Capra, Fritjof. *Bootstrap Theory of Particles.* Revision, Fall/Winter, 1981.

———. *The Tao of Physics.* New York: Bantam, 1976.

——.*The Turning Point: Science, Society and the Rising Culture.* New York: Simon & Schuster, 1976.

——.*The Tao of Physics, 4th edition.* Boston: Shambhala, 2000.

Castaneda, Carlos. *The Teachings of Don Juan.* New York: Ballantine Books, 1968.

Chopra, Deepak. *Quantum Healing.* New York: Bantam, 1989.

Cook, Theodore. *The Curves of Life.* New York: Dover, 1979.

Cooper, R.G. The Performance Impact of Product Innovation. *European Journal of Marketing,* 1984.

Cornell, Ann Weiser. *The Power of Focusing.* New York: MJF Books, 1996.

Costley, D.; K. Downey and M. Blumberg. *"Organizational Climate: the Effects of Human Relations Training."* University Park, PA: Working Paper, Penn. State University, 1973.

Covey, Stephen. *Seven Habits of Highly Effective People.* New York: Simon & Schuster, 1990.

de Broglie, L.; Armand, L, Simon, P.H. *Einstein.* New York: Peebles Press, 1979.

de Chardin, Teilhard. *The Divine Milieu.* New York: Harper and Row, 1960.

d'Entreves, A.P. *Natural Law: An Introduction to Legal Philosophy.* London: Hutchinson & Co., 1951.

Deshpande, Rohit & A. Parasuraman. Organizational Culture and Marketing Effectiveness. In P.F. Anderson & M.J. Ryan (Eds.), *Scientific Method and Marketing.* Chicago, IL: American Marketing Association, 1984.

Deshpande, Rohit, & F.E.Webster, Jr. "Organizational Culture and Marketing: Defining the Research Agenda." *Journal of Marketing,* 1989, 3, 3-15.

Dinsdale, H. "Future Thinking.". *Future Survey.* 1993, 15 (4).

Donnelly, J.H. Jr. & J.M. Ivancevich. "A Methodology for Identifying Innovator Characteristics of New Brand Purchasers." *Journal of Marketing Research*, 1974, 11, 331-334.

Dreher, Dianne. *The Tao of Inner Peace*. New York: Harper Perennial, 1991.

Drucker, Peter F. *Innovation and Entrepreneurship: Practice and Principles*. New York: Harper & Row Publishers, 1985.

____. *The Essential Drucker*. New York: Harper Publications, 2003.

Edwards, P. *Encyclopedia of Philosophy, 1-8*, In P. Edwards (Ed.). New York: Macmillan Publishing Company and The Free Press, 1967.

Einstein, Albert. *Ideas and Opinions*. Sonja Bargmann, trans. New York: Crown Publishers, 1954.

——. *Out of my Later Years*. New York: Philosophical Library, 1950.

Etizone, A. *The Moral Dimension*. New York: The Macmillan Press, 1990.

Evan, W.M., & G. Black. "Innovation in Business Organizations: Some Factors Associated with Success or Failure or staff proposals." *Journal of Business,* 1967, (40), 519-530.

Feldenkreis, S. *Potent Self.* New York: Harper & Row, 1985.

Ferguson, Marilyn. *The Aquarian Conspiracy*. Los Angeles: J.P. Tarcher Inc., 1980.

Fieldler, F.E. *A Theory of Leadership Effectiveness*. New York: McGraw Hill, 1967.

Flower, Robert J. *The Intelligence Cubes...and Their Patterns*. Aurora, CO: Gala Publishing, 1991.

Forrester, J. Critical theory of planning practice. J. Forrester (Ed.) *Critical Theory and Public Life*. Cambridge, MA: MIT. 1985. 202-227.

Friedman, Tom. *The World is Flat*. New York: Farrar Straus & Giroux, 2005.

Fuller, R.B. *Synergetics*. New York: Macmillan Press, 1975

Gardner, Howard. *Multiple Intelligences: The Theory in Practice*. New York: Basic Books, 1993.

Gleick, James. *Chaos*. New York: Penguin Books, 1987.

Globus, G, G. Maxwell. and I. Savodnik. *Consciousness and The Brain*. New York: Plenum, 1976.

Gold, B. "Technological Diffusion in Industry: Research Needs and Shortcomings." *Journal of Industrial Economics,* 1981, *29.* (3), 247-67.

Goldsmith, E. "Supersedence: Its Mythology and Legitimization." *Ecologist.* Sept./Oct., 1981.

Greene, Brian. *The Elegant Universe.* New York: Vintage, 2000.

Gregory, K.L. Native view paradigms: Multiple cultures and Culture Conflicts in Organizations. *Administration Science Quarterly.* September, 1983: 359-76.

Guildford, J.P. Traits of Creativity. In H. Anderson (Ed.), *Creativity and Its Cultivation*. New York: Harper, 1959.

Guthrie, W.K.C. *A History of Greek Philosophy.* Cambridge, England: Cambridge University Press, 1969.

Gawain, Shakti. *Creative Visualization.* New York: Bantam Books, 1985.

Hail, W.K. Strategic Planning, Product Innovation and the Theory of the Firm. *Journal of Business Policy,* 1973 *3*, (3).

Havelock, R.G. *Planning for Innovation.* Ann Arbor, MI: University of Michigan, Center for Research on Utilization of Scientific Knowledge, 1970.

Havener, Cliff. *MEANING: The Secret of Being Alive* Edina, MN: Beaver's Pond Press, 1999.

Hawking, Stephen. *A Brief History of Time*. New York: Bantam Books, 1988.

Hawkins, Gerald .S. *Stonehenge Decoded*. New York: Doubleday, 1964.

Heisenberg, Werner. *Physics & Beyond*. New York: Harper & Row, 1971.

Herzog, A.R. *Subjective Well-Being Among Different Age Groups*. Ann Arbor, MI: Institute of Social Research, 1986.

Hirschman, E.C. "Symbolism and Technology as sources for the generation of Innovations." In Andrew Mitchell (Ed.), *Advances in Consumer Research*. St. Louis, MO, Association for Consumer Research, 1981:9: 537-541.

Hofstadter, D.R. *Godel, Escher, Bach: An Eternal Golden Braid*. New York: Vintage, 1980.

——.*Metamagical Themas: Questing for the Essence of Mind and Pattern*. New York: Bantam Books, 1986.

Hofstede, Geert. *Culture's Consequences*. Beverly Hills, CA: Sage, 1980.

Jung, Carl G. *Modern Man in Search of a Soul*. London: Kegan Paul Trench Trubner: 1933.

——.*Man and His Symbols*. New York: Dell, 1964.

Keepin, Will. "Lifework of David Bohm." www.vision.netau/~apaterson/science/david_bohm.htm.

Keirsey, David and Marilyn Bates. *Please Understand Me*. Del Mar, CA: Prometheus Nemesis Books, 1992.

Kiev, Dr. Ari. *A Strategy for Daily Living*. New York: Simon & Schuster, 1997.

Kuhn, T. *The Structure of Scientific Revolutions*. Chicago, IL: Chicago University Press, 1970.

Langham, Derald G. *Genesa: An Attempt to Develop a Conceptual Model to Synthesize, Synchronize, and Vitalize Man's Interpretation of Universal Phenomena.* Fallbrook, CA: Aero Publishers, 1969.

——. "Genesa." KPFK Radio Interview. Fallbrook, CA: Aero Publishers, 1975.

Laszlo, Ervin. *The System View of the World.* New York: Braziller, 1972.

Lerner, E J. *The Big Bang Never Happened.* New York: Vintage, 1991.

LeShan, L. *The Mystic, the Median and the Physicist.* New York: Viking, 1974.

Lilenfield, R. *The Rise of Systems Theory: An Ideological Analysis.* New York: John Wiley, 1978.

Lovell, Jim and Jeffrey Kluger. *Lost Moon.* New York: Houghton Mifflin, 1994.

Mainzer, Klaus. *Thinking in Complexity.* Berlin, Germany: Springer-Verlag, 1994.

Major British Poets of the Romantic Period. Edited by William Heath. New York: Macmillan, 1973.

Maslow, Abraham. *The Farther Reaches of Human Nature. (2nd ed.)* New York: Viking Press, 1971.

——.*Towards a Psychology of Being.* New York: Viking Press, 1962.

Miller, J.G. *Living Systems.* New York: McGraw Hill, 1978.

Needham, J. *Science and Civilization in China.* Cambridge, England. Cambridge University Press, 1956.

O'Connor J.S and Ian McDermott. *The Art of Systems Thinking.* San Francisco, CA: Thorson's Publishing, 1997.

Pagels, Heinz. *The Cosmic Code.* New York: Simon & Schuster, 1982.

Peale, Norman Vincent. *The Power of Positive Thinking.* Englewood Cliffs, NJ: Prentice-Hall, Inc., 1952.

Peat, F. David and John P. Briggs. *Looking Glass Universe.* New York: Simon & Schuster, 1984.

Peters, Tom. *In Search of Excellence.* New York: Warner Books, 1984.

Plank, M. *The Philosophy of Physics.* New York: Norton Press, 1982.

Pribram, Karl H. *The Holographic Principle.* ???: Freeman Press, 1979.

Prigogine, Ilya. *From Being to Becoming: Time and Complexity in The Physical Sciences.* San Francisco, CA: W.H. Freeman and Co, 1980.

Prigogine, Ilya and I. Stengers. *Order Out of Chaos.* London: Heinman, 1984.

Reich, W. *An Introduction to Orgomy.* New York: Farrar, Straus, and Cudahy, 1960.

Rothman, R.A. *Working: Sociological Perspectives.* New Jersey: Prentice-Hall,1987.

Russell, P. *The Global Brain.* New York: St. Martins Press, 1983.

Sheldrake, Rupert. *Seven Experiments That Could Change the World.* Los Angeles, CA: J.P. Tarcher, 1995.

——.*A New Science of Life: Formative Causation.* Los Angeles, CA: J.P. Tarcher, 1982.

Siegal, Bernie, M.D. *Love, Medicine and Miracles.* New York: Harper Collins, 1990.

Surowiecki, James *The Wisdom of Crowds: Why the Many are Smarter Than the Few and How Collective Wisdom Shapes Business, Economies, Societies, and Nations.* New York: Doubleday, 2004.

Tomkins, P. *Mysteries of the Great Pyramid.* New York: Harper & Row, 1978.

——.*Mysteries of the Mexican Pyramid*. New York: Harper & Row, 1984.

Tracy, L. *The Living Organization—Systems of Behavior*. New York: Praeger Publishers, 1989.

Von Bertalanffy, Ludwig *General Systems Theory*. New York: Braziller, 1968.

Walonick, David S. "A Holographic View of Reality." www.survey-software-solutions.com/walonick/reality.htm.

____. "General Systems Theory." www.survey-software-solutions.com/walonick/systems-theory.htm.

Watzlawick, Paul. *The Language of Change: Elements of Therapeutic Communication*. New York: Basic Books, 1978.

Webster's New World Dictionary of the American Language. 2nd College Edition. Edited by David B. Guralnik. New York and Cleveland: The World Publishing Company. 1970.

Weiner, Norbert. *Cybernetics*. New York: Wiley, 1948.

Weiner, Philip. *Dictionary of The History of Ideas,*. New York: Charles Scribner's Sons, 3,1968.

Wilbur, K. *The Altman Project*. Wheaton, Illinois: Theosophical Publishing, 1980.

Wolf, Fred Alan. *The Dreaming Universe*. New York: Touchstone, 1995.

——.*Star Wave*. New York: Macmillan Press, 1984.

——.*Parallel Universes*. New York: Touchstone, 1990.

Zukav, G. *The Dancing Wu-Li Masters*. New York: Bantam Books, 1979.